Society for Environmental Geochemistry and Health

Lead in Soil

Recommended Guidelines

Edited by

BOBBY G. WIXSON

and

BRIAN E. DAVIES

CRC Press
Taylor & Francis Group
Boca Raton London New York

CRC Press is an imprint of the
Taylor & Francis Group, an **informa** business

SOCIETY FOR ENVIRONMENTAL GEOCHEMISTRY AND HEALTH

"LEAD IN SOIL" TASK FORCE

RECOMMENDED GUIDELINES

Edited by

Bobby G Wixson
College of Sciences
Clemson University
Clemson, South Carolina, U.S.A.

and

Brian E. Davies
Department of Environmental Sciences
University of Bradford
Bradford, West Yorkshire, England

CONTRIBUTORS

Robert L. Bornschein
Rufus L. Chaney
Willard R. Chappell
J.Julian Chisolm Jr
C. Richard Cothern
Brian E. Davies
Betsy T. Kagey
Howard W. Mielke
Albert L. Page
C.D. Strehlow
Rosalind Volpe
Daniel L. Vornberg
Pamela Welbourn
Bobby G. Wixson

Sponsored by

The United States Environmental Protection Agency,
International Lead Zinc Research Organisation, Inc.,
Lead Industries Association,
Society for Environmental Geochemistry and Health,
University of Bradford
and
College of Sciences, Clemson University.

SOCIETY FOR ENVIRONMENTAL GEOCHEMISTRY AND HEALTH "LEAD IN SOIL" TASK FORCE

Robert L. Bornschein, Ph.D.
University of Cincinnati
Dept. of Environmental Health
Cincinnati, OH 45267 U.S.A.

Rufus L. Chaney, Ph.D.
Soil-Microbial Systems Lab, AEQI
USDA, ARS
Beltsville, MD 20725 U.S.A.

Willard R. Chappell, Ph.D.
Environmental Sciences
University of Colorado at Denver
Denver, CO 80217 U.S.A.

J. Julian Chisolm Jr., M.D.
Kennedy Krieger Institute
Lead Clinic
Baltimore, MD 21205 U.S.A.

C. Richard Cothern, Ph.D.
US EPA, 401 M Street. S.W.
Washington, DC 20460 U.S.A.

Brian E. Davies, Ph.D. (Co-chairman)
Department of Environmental Science
University of Bradford
Bradford, West Yorkshire
England BD7 1DP

Ms Betsy T. Kagey, MPH
Empire State College
Albany, NY 12222 U.S.A.

Howard W. Mielke, Ph.D.
College of Pharmacy
Xavier University of Louisiana
New Orleans, LA 70125 U.S.A.

Albert L. Page, Ph.D.
Dept. of Soil & Environmental Science
University of California
Riverside, CA 92521 U.S.A.

C.D. Strethlow, Ph.D.
Dept. of Child Health
Westminster Children's Hospital
London, England SW1P 2NS

Iain Thornton, Ph.D.
Global Environmental Research Centre
Imperial College, Prince Consort Road
London, England SW7 2BP

Rosalind Volpe, Ph.D.
International Lead Zinc Research
Organization, Inc.
Research Triangle Park, NC 27709
U.S.A.

Mr. Dan Vornberg
Smelting Division
The Doe Run Company
Herculaneum, MO 63048 U.S.A.

Pamela Welbourn, Ph.D.
Environmental Centre, Trent University
Peterborough, Ontario, Canada K9J 7B8

Bobby G. Wixson, Ph.D. (Chairman)
College of Sciences
Clemson University
Clemson, SC 29634 U.S.A.

SPECIAL CONSULTANTS

Robert W. Elias, Ph.D.
Trace Metal Geochemistry and Health
Environmental Criteria & Assessment
Office (MD-52), US EPA
Research Triangle Park, NC 27711
U.S.A.

Robert D. Putnam, Ph.D.
Putnam Environmental Services
Research Triangle Park, NC 27709
U.S.A.

ACKNOWLEDGMENTS

The sponsorship of this task force by the Society for Environmental Geochemistry and Health (SEGH), the United States Environmental Protection Agency (U.S. EPA), the International Lead Zinc Research Organization (ILZRO), the Lead Industrie Association (LIA), the University of Bradford, England and Clemson University, South Carolina, USA is gratefully acknowledged. Special thanks must go to Jerome Cole, Lester Grant, John Yoder and Rosalind Volpe for their encouragement and support. Robert Elias from the U. S. EPA gave valuable assistance as a special consultant on modelling and Robert Putnam of Putnam Environmental Services served as an alternate for Rosalind Volpe and their assistance is most appreciated by the task force. The information furnished by Andre Rosenoff and Linda Jennett was most helpful. The SEGH task force would also like to acknowledge and thank all reviewers for their comments and helpful suggestions made on the initial draft report on lead in soil.

The typing and other requirements of first drafts of the report would not have been possible without the patience and good humour of Janet Dillon and Cleve Ann Senn at Clemson University.

Final typing, editing and preparation for publication was done at Bradford. The diagrams were redrawn for publication by Mr S Davidson of the Department of Environmental Science at the University of Bradford. Dr Henrietta Lidiard of the same department undertook final copy editing before sending the report out for SEGH authorisation. Mrs Nirmala Mistry retyped the tables.

The Task Force is most grateful for all the contributions of these organisations and individuals in making this report possible.

NOTICE AND DISCLAIMER

The thoughts and ideas expressed in this report are those of the authors and do not necessarily represent those of the organisations listed.

This report is published as a contribution to scientific debate and had been authorised by the Society for Environmental Geochemistry and Health only for that purpose. The editors and the members of the Task Force have taken every reasonable care to ensure that the report is accurate and up to date. Nonetheless, it is not, nor does it purport to be, an instrument of guidance for the public use. Neither the authors, nor the sponsoring organisations can accept any responsibility for any consequences arising from the use of the report.

First published 1993 by Science Reviews
Taylor & Francis Group
6000 Broken Sound Parkway NW, Suite 300
Boca Raton, FL 33487-2742

Reissued 2018 by CRC Press

©1993 by Taylor & Francis
CRC Press is an imprint of Taylor & Francis Group, an Informa business

No claim to original U.S. Government works

This book contains information obtained from authentic and highly regarded sources. Reasonable efforts have been made to publish reliable data and information, but the author and publisher cannot assume responsibility for the validity of all materials or the consequences of their use. The authors and publishers have attempted to trace the copyright holders of all material reproduced in this publication and apologize to copyright holders if permission to publish in this form has not been obtained. If any copyright material has not been acknowledged please write and let us know so we may rectify in any future reprint.

Except as permitted under U.S. Copyright Law, no part of this book may be reprinted, reproduced, transmitted, or utilized in any form by any electronic, mechanical, or other means, now known or hereafter invented, including photocopying, microfilming, and recording, or in any information storage or retrieval system, without written permission from the publishers.

For permission to photocopy or use material electronically from this work, please access www.copyright.com (http://www.copyright.com/) or contact the Copyright Clearance Center, Inc. (CCC), 222 Rosewood Drive, Danvers, MA 01923, 978-750-8400. CCC is a not-for-profit organiza-tion that provides licenses and registration for a variety of users. For organizations that have been granted a photocopy license by the CCC, a separate system of payment has been arranged.

Trademark Notice: Product or corporate names may be trademarks or registered trademarks, and are used only for identification and explanation without intent to infringe.

Publisher's Note
The publisher has gone to great lengths to ensure the quality of this reprint but points out that some imperfections in the original copies may be apparent.

Disclaimer
The publisher has made every effort to trace copyright holders and welcomes correspondence from those they have been unable to contact.

ISBN 13: 978-1-138-10603-1 (hbk)
ISBN 13: 978-0-203-71154-5 (ebk)

Visit the Taylor & Francis Web site at http://www.taylorandfrancis.com and the CRC Press Web site at http://www.crcpress.com

CONTENTS

I. HOW TO USE THIS REPORT

This report is intended for those concerned with the question of how to evaluate whether they have a potential human health problem arising from local concentrations of lead in soil. They will include public health officials, regulatory agencies and industrial environmental managers. The report has been written so that it is easy to use.

The report has been divided into sections. Section II is an introduction saying why it was written. The introduction is then followed (Section III) by a glossary of definitions used in the report in order to assist the reader in understanding the meaning of specialised terms.

Section IV is a protocol or logic format called a *phased action* plan. This is used in a step wise progression through six major areas which are shown in Figure 1. These are summarised below. Each step contains a reference to subsequent chapters in the report where the background information and other details are provided.

The first steps (A to D) concentrate on assessing whether there is a *problem* with an observed concentration of lead in soil.

The first area, on the left, is headed **Problem.** Many proven health problems with lead have arisen from some chance discovery of a high lead value or from a suspicion that there might be lead contamination locally or a clinical observation of lead problems in animals.

A. *Unplanned elevated soil lead*

A soil lead concentration suggestive of a possible health problem but not derived from a systematic, quality controlled survey programme.

B. *Unplanned elevated blood lead*

A concentration of lead in blood (PbB) which might arise from some clinical investigation which was not originally designed to investigate a specific lead in blood problem.

C. *Unplanned discovery of lead in animal or plant tissues*

Some indication of lead toxicity in domestic animals, wildlife or plants.

D. *Anticipated potential lead in soil problems*

Land that was previously in industrial use and might therefore be suspect with respect to lead contamination.

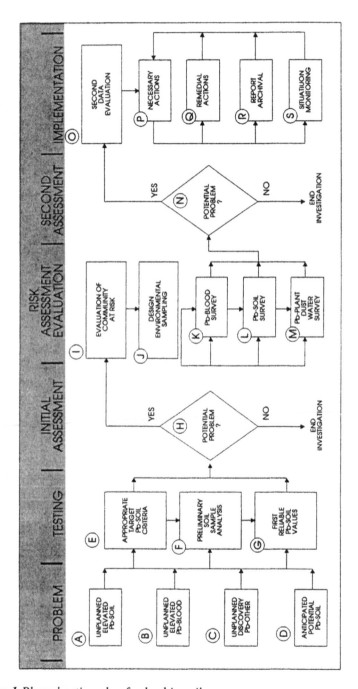

Figure 1 *Phased action plan for lead in soil*

If any of these problems are found to be present, then the flow of the plan moves to the second area entitled *Testing* which requires the use of:

E.   Appropriate target lead in soil criteria

This is a computable value based on specified criteria. The relationship used is explained in Section VII. Supplement I contains a simple computer program to aid in the calculation together with a series of worked examples.

F.   Preliminary soil sampling and analysis

A preliminary systematic and rigorous survey must be carried out to characterise soil lead contents in the area under investigation. Recommended methods for soil sampling and analysis are to be found in the back of the report as a supplemental section.

G.   First reliable soil lead values

These are obtained from the preliminary soil sampling and analysis. They allow one to characterise the area for the next level of the decision making process, *i.e.* Step H.

The third area in the action plan requires an *Initial Assessment* of the *Potential Problem* (Step H) based on the data obtained and the target levels chosen. A decision is now made either to *end the investigation* or to proceed to the *Risk Assessment Evaluation* area.

The *risk assessment* evaluation contains a series of steps to be made, namely, I to M:

I.   Evaluation of the community at risk

Here it is necessary to determine and, where possible quantify. other contributing factors such as the number and age of the population at risk or land use. These may well reveal the need for comprehensive environmental sampling.

The supplemental section of the report may again be consulted for sampling and analytical methods appropriate for use on a site specific basis.

J.   Design environmental sampling

A general description of the project is needed including necessary details associated with time tables and tasks, use of data, project organisation and responsible individuals.

K.   A blood lead survey

This needs to be designed and performed by appropriately trained medical personnel utilising a laboratory with an acceptable quality control programme.

L.   A soil lead survey

This needs to be made for those areas indicated by the risk assessment and environmental design. Details on sampling and analysis are again to be found in the special supplemental sections of the report dealing with sampling methods.

M. Surveys of lead in dust, vegetation and water

These are also to be conducted by methods as illustrated in the supplemental sampling and analysis section of the report

The *Second Assessment* decision (Step N) is then based on the information gained during the risk assessment evaluation. At this time a decision must again be made on either to *end the investigation* or to proceed to the *implementation* (risk management) stage (Steps O to S).

The implementation process requires the consideration of:

O. A second data evaluation

Available information must now be assessed in terms of the financial resources available for the various actions which may need to be taken.

P. Necessary actions

A site specific risk management decision process will indicate what to do next. Details concerning how to develop a risk management strategy are noted in the section of the report dealing with this subject.

Q. Remedial actions

These may be prescribed but will need more scrutiny. See the risk management section of the report.

R. Report archival

All data collected and evaluated should be stored in an appropriate location for possible future use.

S. Situation monitoring

The problem site should be monitored in a systematic manner to determine the efficacy of the remedial actions.

Details on the various steps described above are to be found in the Phased Action Plan section (IV) of the report.

Section V provides a concise overview of lead in soil. Information is given about the concentrations of lead found generally in normal soils together with an account of soil formation.

The bioavailability of lead (Section VI) has now become a major area of concern when evaluating the effects of lead in regard to human health. This section describes the factors that influence the biovailability of lead in regard to chemical composition, particle size and nutrition factors.

The Health section of the report (Section VII) discusses the population groups at risk, definitions of acceptable blood lead concentrations, other sources of lead and evaluations of appropriate case studies that serve as examples. The use of health criteria in deriving the target soil/dust lead guideline concentration model should be examined carefully since this is the model used to determine the appropriate target soil lead/blood lead action level in the Phased Action Plan. Since a single value is at best of limited use or, at worst, misleading, the examples given will help in understanding how the model and number range derived may be applied to different situations.

Descriptive information relating to the overall analysis of risk as associated with the suggested soil lead guidance is presented in Section VIII, the risk management section of the report. Possible remedial actions and associated cost considerations will be of interest to anyone concerned with implementing necessary actions to protect human health.

Acknowledgements (IX) are followed by the references (X) quoted in each section and arranged by chapter.

The Supplements contain practical advice. Supplement 1 is a computer program in BASIC for use in a PC to compute target soil lead concentrations. Recommended methods presently in use for soil sampling and analysis are also presented in the report as supplements. These methods and sampling designs should be of value for use in various site specific evaluations that require quality control and the use of soil data for further evaluation and use in the Phased Action Plan matrix.

II. INTRODUCTION

At present there is a lack of well founded guidelines in the United States and other countries for evaluating concentrations of lead in soil in relation to a possible effect on human health. This has contributed to confusion among regulatory agencies, industries, public health officials, the medical community and citizens concerned with evaluating or remedying lead contaminated soils. Public health officials and the medical community are often expected to judge the health effects and risks from lead exposure. As a result they must also decide the soil lead concentration that should be used as the basis for requiring remedial action at contaminated or hazardous waste sites.

The need for science based guidelines was emphasised in a special session of the 1987 Trace Substances in Environmental Health Conference held in Columbia, Missouri, U.S.A. following a key note presentation on "Lead in Soil: How Clean is Clean?" by Davies and Wixson (1986). As a result of the questions raised and the continuing scientific concern expressed, a special conference on "Lead in Soil: Issues and Guidelines" was then held in Chapel Hill, North Carolina in March 1988 under the sponsorship of the Society for Environmental Geochemistry and Health (SEGH), the United States Environmental Protection Agency (EPA), the International Lead Zinc Research Organization (ILZRO), the Lead Industries Association (LIA) and Clemson University, South Carolina. At this meeting over thirty scientific papers were presented which summarised pertinent scientific data and reported previous and on-going lead in soil case studies.The guidelines or scientific approaches used in various countries throughout the world were also reported and discussed. The papers presented at this special conference have been published by *Environmental Geochemistry and Health* as a special proceedings entitled "Lead in Soil: Issues and Guidelines", edited by Davies and Wixson (1988).

The conference featured panel and audience discussion on suggestions for possible approaches to be used in the development of U. S. guidelines for lead in soil. A "phased-action plan" approach was proposed by Wixson (1988) and accepted by the participants together with a request that a Society for Environmental Geochemistry and Health (SEGH) task force be formed to evaluate further the conference findings. The need for a report was recognised that would recommend guidelines for lead in soil based on a critical selection of the best available scientific data and knowledge.

In June 1988, a status report entitled "Lead in Soil: Issues and Guidelines Conference Summary" was presented by Wixson (1989) at the Trace Substances in Environmental Health Conference held in St Louis, Missouri. A special task force was then approved to study and report on lead in soil under the auspices of the SEGH. The task force has been supported by the U.S. Environmental Protection Agency (EPA), International Lead Zinc Research Organization (ILZRO), the Lead Industries Association (LIA) and Clemson University.

The "Lead in Soil" task force was drawn from SEGH members and represented a balance of established and reputable scientists from regulatory agencies, industries, the medical profession, public health authorities and universities: all are active workers in this subject. The multidisciplinary membership of the task force enabled it to integrate reports and data from across a wide spectrum of the scientific literature. The intense, interactive nature of the task force meetings provided opportunities for a highly focused selection of relevant information.

The major objective of the task force was to produce a report utilising a flexible matrix approach or *phased action plan* to evaluate the evidence and problems in its interpretation and, hence, to make recommendations for guidelines to appraise lead concentrations in soil.

The SEGH task force held a succession of meetings from which emerged this report which contains an evaluation procedure firmly rooted in good scientific data. A phased-action plan was coupled with target soil lead concentrations derived through a model relating blood lead and soil lead concentrations. Such a plan allows for the combined influences of soil and other sources of lead on blood lead concentration. Thus, the model offers flexibility for the user to select appropriate target levels of blood lead concentrations while allowing for a variety of environmental situations or regulatory criteria.

III. GLOSSARY AND DEFINITIONS

Scientists use some words and terms in a very specialised and closely defined manner. Sometimes these words are used in everyday speech but with a wider or looser meaning. Also, words have been invented to provide for a specific communication need. The following glossary of such words is provided for the reader of this report.

BACKGROUND LEAD CONCENTRATION

The concentration of lead in soil at sufficient distances from known mobile or point sources of contamination such that it is representative of typical soils for the region in question.

BIOAVAILABILTY

For a given substance, different physical/chemical forms have different availabilities to, and therefore different effects upon, living organisms. Rarely is all the substance that is ingested or otherwise taken into an organism, absorbed by that organism. This means that a value for total lead in a real environmental sample is almost always an overestimate of the amount of lead that is available and that will be absorbed. However, in the absence of reliable information concerning the form of lead in a sample, one should assume 100% availability. Moreover, at the present time only total lead can be determined accurately and precisely.

For scientific purposes, lead acetate, a water soluble lead salt, is used as the standard for 100% availability of dietary lead. When equal amounts of soil lead, or large particle lead sulphide, are added to the same test diet, the absorbed lead is lower than that of lead acetate. This is due to the physical and chemical properties of these lead sources. The standard diet for bioavailabilty testing is a casein-sucrose (AIN 78) purified diet, because high fibre or phytate in diets significantly lowers the fraction of dietary lead that is absorbed.

CONTAMINATION

Soils which are formed on similar parent materials and which have similar arrangements of horizons, *i.e.*, they have formed under similar environmental conditions and are of similar age, are grouped together by pedologists in a soil *series*. There is evidence to suggest that the trace element concentrations, including lead, for a large number of samples from a given soil series in the natural state can be described, in statistical probability terms, as having an upper and lower limit. The frequency distribution of lead values from a large number of samples is usually positively skewed but this skewness can be minimised, *i.e.* the population normalised, by transforming the values to their $\log_{10}$ equivalents. The (geometric) mean value and the 95% or 99% probability ranges can then be calculated. Values for other samples apparently from the same series that do not lie within the range are described as anomalous and form a geochemical anomaly. If, from other evidence, it is deduced that the anomaly has been caused by anthropogenic activity, the geochemical anomaly becomes a neoanomaly and the soil is regarded as contaminated. Thus, a contaminated soil is one having a lead content larger than that expected from the pedological nature of the soil.

In practice, it may not be possible to identify a suite of uncontaminated soils for the purposes of comparison. Statistical techniques do exist for extracting the basal population. But this population is then better described as 'background' rather than natural and uncontaminated.

DIRT

An ill-defined word which includes soil and dust and should not be used.

DUST

Loose organo-mineral material lying as a thin veneer on a solid support. It is differentiated from soil in that soil is presumed to have formed by pedogenesis from its rock base whereas dust has simply settled on its base. Dusts should be described as street, kerb-side, settled-house or air-borne.

GARDENS OR YARDS

Excluding back-to-back terraced houses in industrial England most houses have associated with and adjacent to them a patch of open ground. In Great Britain part or all of this may be paved at the rear and is called the 'back yard' or when at the front, the 'forecourt'. Unpaved areas are used for growing grass (lawn), decorative plants or vegetables. All these unpaved areas are called 'garden'. In North America open areas are generally described as 'yard' and 'garden' will ordinarily imply a vegetable or flower plot.

GRADE

In the USA successive age levels in public schools are described in terms of grades, where first grade applies to the youngest entrants to elementary school at age 6. Younger children attend *kindergarten*. But in Britain, where children enter *primary school* at age 5 (nursery school before that) the term grade is not used. In most areas (but not all) children at 11 go to a secondary school and are *first formers*. Pupils may leave school at age 16 (fifth formers), or continue in school to age eighteen as *sixth formers*. In many schools these pupils have their own facilities in a *sixth form college*. These terms are often adopted in other English-speaking countries. *Forms* or *grades* provide a short-hand to describe a particular group of pupils or students: thus, fourth formers, or grades K through 12.

HUMUS

Dead or senescent organic material is food for other organisms which, in turn, are themselves consumed. The end product is very different chemically from biological tissue, provides highly reactive surfaces, is relatively persistent in soil (decades or centuries compared with months or years for fresh litter) and is one of the features which distinguishes soil from simple weathered regolith. Humus forms by humification and decays to inorganic solutes and gases by mineralisation.

LITTER

Grass, dead leaves, decaying leaves, on the surface of a yard (garden) soil. Material at the beginning of the humification process.

NEIGHBOURHOOD

A group of homes having similar external appearances and normally residents with similar socio-economic status, extending over an area of 16 – 25 square blocks (0.5 – 1 square mile), or equivalent to a medium density U. S. census tract.

PEDOGENESIS

The formation of soil from rock (both hard rock and soft sediments) by a combination of chemical decomposition, physical break down and incorporation of humified soil organic matter. The latter process, involving biological activity, differentiates pedogenesis from weathering.

PICA

An abnormal craving for certain unnatural substances such as clay, chalk, soil, dust, paint chips or other non-edible materials.

PLUMBOSOLVENCY

The dissolution of lead from pipes, solder or lead-lined tanks used in domestic plumbing systems. The problem is most acute for waters of low pH and low calcium content and after such water has remained stagnant in contact with the lead for several hours. In recent years in Britain many water utilities have increased the hardness of piped water to minimise the problem and there has been an extensive programme of removing lead pipes from older houses.

POLLUTION

A term which has been expressed in many different ways. Wanielsta *et al.* (1984) have described pollution as "that which modifies the environment such that its use is affected." Warren (1971) describes pollution as a word that has social significance and suggests that it is much used by individuals or groups to take on different meanings to stimulate legislation or pass laws.

The report by the National Research Council Committee on Pollution (1966) starts off with "Pollution is an undesirable change in the physical, chemical, or biological characteristics of our air, land, and water that may or will harmfully affect human life or that of other desirable species, our industrial processes, living conditions, and cultural assets, or that may or will waste or deteriorate our raw material resources".

This means that a soil may be described as contaminated yet the concentration of lead in that soil may not be large enough to cause any deleterious biological effects. But where the lead concentration does rise to a value whereby a specific organism is adversely affected then the soil may be described as polluted. Since pollution is an anthropogenic process then a soil should not be described as polluted where the lead content has reached injurious levels through natural processes.

Based on these various explanations and adapting soil from the water usage terms of Warren (1971), this report should simplistically define soil pollution as "any impairment of the suitability of soil for any of its actual or beneficial uses, by man-caused changes."

PROPERTY

In North America the word is used generally to describe the plot of land upon which a building is situated and belonging to that building through freehold or leasehold. The term is in use in Great Britain, but less widely, and may be restricted to the building itself.

QUALITY ASSESSMENT (QAS)

The overall system of activities to provide assurance that the quality control task is being performed effectively. Quality assessment involves a continuing evaluation of performance of the production system and the products produced.

QUALITY ASSURANCE (QA)

A system of activities, the purpose of which is to provide the producers and users of a product or service with the assurance that it meets defined standards of quality with a stated level of confidence. The QA system includes the separate but coordinated activities of Quality control and Quality assessment.

QUALITY CONTROL (QC)

The overall system of activities designed to control the quality of a product or service so that it meets the needs of users.

SCHOOL YARD

The area adjacent to a school where children play.

SOIL

The loose and weathered veneer of material overlying and merging with rock from which it has normally formed. Agronomists restrict the meaning of soil to the weathered material in which plants grow and distinguish pedogenetic processes from geological weathering. Engineers extend the term to all loose surface materials. The agronomic usage is adopted here.

Yard or garden soil

The soil found in gardens or yards. Surface soil is the top 1 – 2 inches (2 – 5 cm) which contains the roots of most garden plants, while subsurface soil lies below 6 inches or 15 cm. Litter includes grass, leaves *etc.* lying on the soil. Any surface litter accumulation is regarded as 'above surface material'.

Subsurface yard or garden soil

The soil located in yards (gardens) at a depth of 2 – 6 inches (5 – 15 cm) below the litter level.

TRIGGER VALUE

The term used by the British Department of the Environment (Simms and Beckett, 1986) for concentration of lead in soil which, for the particular area and set of conditions under consideration, indicates the need to evaluate the risk to human health from the soil lead and, if necessary, leads to some kind of remedial action on or limitation of use of the land. There is no absolute value of lead in soil that can be used as a trigger.

IV. PHASED ACTION PLAN

This chapter is concerned with the flow diagram of the phased action plan, shown in Figure 1.   Discussion is restricted to a concise description of each step as noted in the sequence A to S contained in Figure 1.

Problem (A to D)

The four steps A, B, C and D are grouped together in the *problem* area. Experience shows that many health problems involving lead originate from some chance discovery or from some speculation. An individual may have a soil analysed and an apparently high lead content is reported. Land is purchased with a view to redevelopment and some knowledgeable local citizen draws attention to its past industrial history. Any of the outcomes from the four boxes may trigger a health alert.

Step A: Unplanned Discovery of Elevated Soil Lead

(See Section V)

In our definitions, we drew a careful distinction between soil and dust (and advised against using the word dirt). In practice, the distinction may be one without a difference. Much of the public health literature does not draw the essential distinction between soil and dust. In this report a relaxed attitude is adopted and dust and soil are regarded as crudely synonymous where this seems apropriate.This report is primarily concerned with the evaluation and action resulting from lead data derived from a systematic sampling of soil. However, an analysis for lead may be carried out on any sample of soil as a result of an individual's curiosity or suspicion.   This is therefore termed an unplanned discovery. Unplanned discoveries may also emerge from soil surveys carried out for other purposes. The analytical value from such a process cannot be accepted uncritically; it should therefore be regarded as a crude value. Soil sampling and analysis for lead needs specialised knowledge, skills and experience. It is prudent to be sceptical about lead contents reported by individuals, albeit skilled professionals in other areas, who do not have this specialist background. Go to Step E.

Step B: Unplanned Discovery of Elevated Blood Lead

(See Section VII).

Concern about lead in soils and dusts might initially be raised as the result of the accidental discovery of one or more elevated blood lead values which might arise from clinical investigations in the area in question. Any elevated blood lead found in this way must be confirmed by repeat sampling and analysis and evaluated with respect to the health criteria adopted for the particular population.

Moreover, as in the case of soil work, sampling and analysis of blood for lead must be undertaken by skilled workers if the reported concentrations are to be regarded as reliable. Readers in countries where there are no regulations or laws concerning limits for lead in blood should refer to Section VII for suggestions for concentrations to be used in deciding whether a particular blood lead level is elevated.If the circumstances of the individual(s) tested, especially a group of children living in the same neighbourhood, suggest that soils and dusts could be a factor contributing to their apparently unacceptable blood lead concentrations, then consideration should be given to proceeding to Box G for obtaining the first reliable soil lead values. Otherwise, go to Step E.

Step C: Unplanned Discovery of Elevated Lead in Other Media

This is the discovery of elevated lead values in animal or plant tissues, or some indication of a lead toxicity problem in livestock, domestic animals or wildlife, which imply the possibility that soil lead might be involved in the cause of the raised lead value. Again, it is important to recognise the need for skill and experience when analysing other biological or environmental materials for lead. As a general and initial guide, a plant tissue content of greater than 10 g Pb g^{-1} dry matter would be regarded as possibly elevated.

Step D: Identification of a Potential Lead in Soil Problem

(See Section V).

Situations can sometimes be recognised where there are reasons to suspect soil lead problems exist. The suspicion may arise from considering the history of land use (a smelter once stood there, perhaps), possible soil contamination from exterior paint on older houses or structures, or from recognising conditions of lead toxicity or excessive accumulation in biota. The further use of such land would lead immediately to a preliminary soil sampling (Box F).

Testing

The flow in Figure 1 has now moved to the second area. Enough information is on hand to proceed to an *a priori* assumption that a local health problem with respect to soil lead may exist. Bearing in mind the important qualification that soil lead contents reported by any laboratory not usually engaged in this kind of work cannot be regarded as necessarily valid it is necessary to revisit the locality in question with a view to taking further soil samples.

Step E: Appropriate Target Soil Lead Criteria

(See Sections V, VI and VII).

Despite the hesitation that must be present when interpreting the soil lead value emerging from Step A it needs to be judged. A common faulty conclusion is to look up a value for "normal " soil lead and then express concern if the observed soil lead is several times greater than the book value. Reference books usually cite normal values as anywhere from 15 to 50 µg Pb g^{-1} soil yet a value an order of magnitude greater does not necessarily entail concern. Proposing a single value guideline for an upper concentration of lead in soil to protect young children was considered unrealistic for several reasons. Various blood lead concentrations are used as health standards around the world and these change as the health effects of lead are reinterpreted. There has been a trend both to an observed decrease of blood levels in the general population and to a change in standards downwards for acceptable blood lead concentrations. Criteria adopted at one time and place may not be suitable for all situations. The individual evaluating the blood lead value may be unaware of changes in legal or advisory limits.

The environment of the population at risk can vary widely. People may be exposed to lead in urban dusts derived from automotive emissions and leaded paints or to soil and dust contaminated by smelter emissions. Lead contamination is often widespread in old lead mining areas. Waste disposal sites can cause metal contamination locally. The population at risk can itself vary from situations where there is a high proportion of young children to a retirement home for the elderly. In other words, the interpretation is complex and interactive.

Because of these considerations, the soil/dust 'guideline' is proposed as a relationship or formula. This allows adjustment for a variety of environmental situations and regulatory criteria. A number of recent papers have described modelling techniques applicable to multiple source exposure to lead and these are discussed in the Health section (VII). Alternate models or means of assessing the blood lead/soil lead relationship can, of course, be used in place of the relationship used here. It is important to emphasise that any such alternate relationship must be used within the framework of this document so that a proper assessment of the situation and any risks involved can be made.

In the model we have developed, blood lead concentration is equated to a baseline level plus an increment resulting from exposure to soil or dust lead. The model takes account of the chosen blood lead guideline or target concentrations and the degree of protection required in the population. The slope of the blood lead – soil lead relationship used in calculating an increase in blood lead over a baseline value, and hence the soil guideline, can vary depending on a variety of factors. Thus, the response can be adjusted for a given situation and modified as more data become available. The relationship derived is as follows:

$$S = \left(\frac{\dfrac{T}{G^n} - B}{\delta} \right) 1,000$$

where:

S is the soil or dust guideline, a geometric mean concentration in g Pb per gram of dust (i.e., ppm). T is the blood lead guideline or target concentration, in g Pb dl^{-1} whole blood. G is the geometric standard deviation of the blood lead distribution, typically in the range of 1.3 to 1.5. B is the background or baseline blood lead concentration in the population from sources other than soil and dust. Data from an appropriate control group would be ideal – a group matched not only for population characteristics, but also for similar lead exposure from all sources except soil and dust. If there are appreciable contributions from other sources such as smelter emissions or leaded paint, these must be measured or estimated for addition to the baseline value. If these data are not readily available, any proposed investigation should evaluate the contributions from other suspected sources. **n** is the number of standard deviations corresponding to the degree of protection required for the population at risk, and would normally follow from the way in which the blood lead guideline T was defined; *e.g.* that 95% of the population should have blood lead concentrations less that 20 g dl^{-1} . Parameter **n** can be obtained from standard statistical tables, and some representative values are given in Table 1. for different percentages of the population desired to be below the target blood lead concentration. δ **(delta)** is the slope or response of the blood lead - soil (dust) lead relationship and has the units of g Pb dl^{-1} blood increase per 1000 μg g^{-1} increment of soil or dust lead.

Step F: Preliminary Soil Sampling and Analysis.

(See supplementary material).

If the information arising from steps A, B, C or D taken together, if appropriate, with the calculation made in Step E, support the view that a health problem might exist arising from contaminated soil then a preliminary, systematic and careful soil investigation must be made. If preliminary soil sampling is required, it is necessary to delimit the area to be sampled (the 'neighbourhood') and design an appropriate sampling and analysis protocol (including quality control procedures). To evaluate the extent of a possible soil lead problem, the preliminary sampling protocol is designed to characterise typical soil lead levels in a neighbourhood, rather than for individual houses, or particular spot locations around a house. A case example is given in the supplement. After the preliminary study is done, a more detailed sampling protocol would include sampling around the house. A soil sample will normally be taken for the house, generally in the centre of the open space behind the built structure as well as other samples (see Step L). In the preliminary and follow up sampling and analysis, as at all stages of the investigation, it is necessary to take precautions against contamination or bias of samples, and to prescribe and adhere to stringent quality assurance (QA) and quality control (QC).

Step G: First Reliable Characterisation of Soil Lead Values

The work done in Step F provides data for soil which can be regarded as reliable in the sense of having been derived from good quality sampling and good quality analysis. A relatively simple statistical treatment of the data will result in the characterisation of the area in terms of the 'first reliable soil lead value'. This is the value that is used for the next level of decision making (Box H in Figure 1).

Table 1. *Values of **n**, the number of standard deviations corresponding to the degree of protection needed for a population at risk.*

Percentage of Population T	Approximate value of n
50	0[*]
95	1.64
98	2.05
99	2.32
99.9	3.04

[*]Target is mean

Initial assessment
Step H: Potential Problem?

An initial assessment can now take place. The soil lead information which is now available in the investigation, and upon which a decision can be made, is technically superior to that which was used to enter Box E. The soil may appear to be contaminated but from an assessment of the mean, median, range or selected quartiles there is apparently no significant health risk (see Step E). If this is so the investigation ends. Otherwise the investigation proceeds to Step I.

Risk Assessment Evaluation

The decision has been made that, assuming a simple and direct relationship between the lead content of soil and that of blood, there is a potential health risk. But the real life situation is far more complex. A local plumbosolvency problem would exacerbate the problem as would the prevalence of flaking, leaded paint in a neighbourhood with many small children. In contrast, a high value for soil lead where the area of soil surface was relatively very small or where the residents were few and elderly, would be far less of a problem. The steps I to M in this area are designed to provide a more realistic evaluation of the problem and lead to another yes/no decision branch in Step N.

Step I: Evaluation of Community at Risk.
(See Sections VI, VII and VIII).

Evaluation is based on an exposure assessment and, in particular, the development of a relationship between blood lead concentrations and the contents of lead in soil where soil is considered to be one possible source of exposure.

The objective of this analysis is to propose a suggested guidance for the relationship between lead in soil and the results of blood lead levels. This relationship forms part of the exposure assessment. Other parts of the exposure assessment include contributions to blood lead from dust, water, food, paint and other media or sources. Among the numerous technical and non-technical aspects that need to be considered are the number and age of the exposed population. If the area of concern contains children living in low income housing or is one frequented by children, such as school yards or playgrounds, the pollution hazard is far more significant than if it contains commercial buildings such as factories or warehouses or if children are likely not to represent a significant proportion of the population, *e.g.*, retirement communities.

The present and probable future land use needs to be considered in deciding whether and what kind of remedial effort is required. Overall risk assessment can be achieved on a site-specific or case-by-case basis. If the major exposure route is from the soil then the guidance suggested in this report may be used directly to determine clean up levels.

The nature of further sampling will depend upon the specific site and its set of conditions and, clearly, it may not be possible to sample all media. For example, in an area where development has not begun there will be no domestic plumbing from which drinking water supplies are derived and water analysis is not relevant. Nor can there be paint to scrape from existing structures. It may even be necessary to estimate the potential exposure to lead from other sources, since this stage leads to a risk analysis, for which other sources should be considered. Even though the purpose of the present procedure is to evaluate the impact of soil as a source of lead, judgment cannot be made without considering potentially confounding factors in as quantitative a manner as possible.

In contrast with the preliminary soil sampling (Box F), which is conducted without any measurement of the effect on the human population, this step will usually include a comprehensive sampling, including blood from most households that have children. It is therefore recommended that the appropriately comprehensive sampling of other media be carried out at the same time as the blood lead survey (if one is conducted). Soil, paint, dust and water should be sampled at every home that provides a blood sample. This avoids any nuisance that might arise from revisiting and intruding on the home. The decision can then be made later whether or not to analyse all these other samples for lead.

For the detailed sampling, lead should be determined as follows:
1. Blood lead in children 6 months to 6 years of age.
2. Soil lead for surface yard or garden samples.
3. House dust.
4. Interior and exterior paint.
5. Drinking water.
6. Street dust.
7. Air.

The detailed procedures for sampling soil, dust, paint and air are described in the supplemental sections of the report. The purpose of this second level of sampling is to characterise the soil lead more accurately and in more detail, as well as to evaluate other sources of lead.

Step J: Design of Environmental Sampling

The U.S. Environmental Protection Agency has outlined a number of items which must be considered for inclusion in a QA project plan (U.S. Environmental Protection Agency, 1979). Those pertinent to the collection of data include: (a) project description, organisation, and responsibility, (b) QA objectives for the measurement of data in terms of precision, accuracy, completeness, representativeness and comparability, (c) sampling procedures and custody, (d) calibration procedures and frequency, (e) analytical procedures, (f) data reduction, validation and reporting, (g) internal and external quality control checks, (h) performance and systems audits, (i) specific routine procedures used to access data precision, accuracy and completeness, (j) corrective action, (k) QA reports to management. A brief description of these items is given below. Additional information can be obtained from supplemental materials at the end of this report. The environmental sampling programme should provide a general description of the project including experimental design. It can be brief but should be sufficiently detailed to allow those responsible for reviewing and approving the programme to complete their task. It should include a timetable for the initiation and completion of tasks within the programme, a statement of the purpose for which the project is planned, and the intended use of the data. The plan should show project organisation and line of authority. Individuals responsible for ensuring the collection of valid data and for the assessment of measurement systems for their precision and accuracy should be identified. A person responsible for carrying out the provisions of the QA plan, a QA officer/manager, should be appointed and identified. All project personnel with responsibility for the quality of the data should receive a copy of the QA project plan and be aware of its contents.

Step K: Blood Survey.

Any detailed blood survey must take account of several important considerations before it is started. The health criteria against which the survey results are to be judged should be agreed upon during the planning stages and procedures established for providing appropriate medical and environmental follow-up of any individuals whose blood lead content exceeds the projected guideline value. This would be necessary whatever the decision taken about any action concerning the area and population in question. Blood samples should be taken by personnel properly trained to avoid sample contamination and using demonstrably low-lead materials. The blood should be analysed only by a laboratory experienced in routine blood lead analysis and whose quality assurance programme includes acceptable performance in an external quality control scheme. The individual results should be made available to the participants along with an interpretation of the survey findings. Any individuals suspected of being unduly affected must be referred to the appropriate physician or hospital for further investigations.

Step L: Soil Lead Survey.

For detailed soil sampling an extensive set of samples should be taken. These should represent the side of the house, the centre of the front garden (yard), if one exists, as well as a three point transect across the back of the yard (garden). In these cases, each sample will be analysed separately instead of bulking them together. This will provide more detailed information on the extent, location and source of the contamination. It is probable, because of the shedding of paint from buildings and impaction of lead aerosols on buildings, that the absolute value of the samples taken from the sides of buildings will be higher than those collected within 1 metre of roads or in yards of the preliminary soil sampling.

For schoolyards, public playgrounds, parks or other amenity areas, the detailed soil sampling will utilise a grid sampling pattern in order to map the extent of the contamination. A sample should be taken from the intersection of each of the grid lines. For vacant and industrial areas or for agricultural land, the grid system should also be used. Again, the sample from each grid intersection must be analysed separately although that sample may be made up of individual soil cores. Detailed sampling programmes containing illustrations on techniques, number and location of samples and other pertinent information are found in the supplemental section of the report.

Step M: Surveys of Lead in Dusts, Plants and Waters.

Sampling protocols for plant, water, dust, interior and exterior paint, and food are covered in the supplemental materials on sampling.

Second Assessment
Step N: Potential Problem?

(See Sections VI, VII and VIII).
Based on the evaluation of the information gained during the various steps of the risk assessment evaluation, the decision should be made either to end the investigation or to proceed to the Implementation Stage. This starts with Step O and is concerned with a second data evaluation.

Implementation
Step O: Second Data Evaluation

(See Section VIII).
The nature and scope of any remedial actions make it important to ascertain the availability of financial resources. Depending on the specific site location, size and uses, the contaminated site may be eligible for public sector assistance (*e.g.*, in the U.S.A., community, state and federal resources) which may supplement any private sector funds. Without this inventory of resource availability, inadequate or unrealistic remedial actions may be proposed.

The costs involved with the physical cleanup of the soil are not always the only ones incurred during remedial actions. Legal liabilities may be created through either taking or not taking action. These play a significant role in determining the scope of the remedial action plan. The costs of monitoring a site after the cleanup has been completed should also be included when estimating the total cost of remedial action.

Step P: Necessary Actions

(See Section VIII).
Based on the findings of the second data evaluation made in Step O, a risk assessment/management plan for the specific site needs to be considered. Factors involved in the risk management decision process include economic, legal, political and social aspects. The uncertainties and non-technical issues concerned with the risk assessment and management are presented in the supplemental materials at the end of the report.

Step Q: Remedial Actions

If the risk assessment/management plan prescribes remedial actions then a number of issues must be considered. These include economic and financial considerations and legal liabilities concerned with taking or not taking action; various types of soil treatment to reduce potential health risks, community education and behaviour modification. More specific information on these issues and recommendations for cost effective methods are contained in the section on risk management and remedial actions.

Step R: Report Archival

All data collected and evaluations made throughout the various steps of the phased action plan for lead in soil need to be retained in an appropriate public domain archive (*e.g.*, the appropriate agency office designated by federal or state law in the USA or more generally in a library or computer data base). The reports should be available for later review or use if there is a change in the usage of a specific site or the population at risk. If for any reason it becomes necessary to continue additional evaluations of a site then the data collected, evaluated and decisions reached earlier will serve as a basis for the later assessments with a consequent saving in time and money. It is therefore imperative that good public domain records be maintained for possible future use.

Step S: Situation Monitoring

After remedial action has been completed, the site must be monitored to ensure that the cleanup actions remain effective. The scale, duration, cost of monitoring and record keeping depends on the type of action taken and should be included in the project budget. Finally, after a further passage of time, the action should again be evaluated and recorded for future reference.

V. LEAD IN SOIL

Average Contents of Lead in Rocks

Ordinarily, lead is a trace element (% by weight) in rocks and soils. But lead also has a strong affinity for sulphur, *i.e.* it is chalcophilic and it therefore concentrates in any separating sulphur phase. The occurrence of such phases may make a rock richer in lead than might otherwise be expected. The major ore mineral is the sulphide (galena, PbS).

There is general agreement that the average crustal rock contains approximately 16 mg Pb kg^{-1}. Nriagu (1978) has reported the lead contents of typical igneous rocks (Table 2). Although 95% of crustal rocks are of igneous origin, sedimentary rocks are spread over the igneous basement and account for 75% of surface exposures; they are therefore the most widespread soil parent materials. The most common sedimentary rocks are shales and mudstones (80%) which have an average lead content of 23 g g^{-1}. Table 3 illustrates the lead content of typical sedimentary rocks. It should be noted that 'black shales' are rich in pyrite (FeS_2) and organic matter and therefore may contain much higher amounts of lead.

Natural, Background and Baseline Contents of Lead in Soil

Estimations of lead in uncontaminated soils vary and only in very recent years has it become widely recognised that such estimations must be based on a statistical appraisal of good quality data. All early estimates tend to be very subjective summaries of data weighted and distorted by sampling in areas where lead concentrations were thought to be unusually high. Hence loose statements have typically been made that the 'normal' lead content of soil ranges from 10-200 g g^{-1}. Table 4 provides data from recent publications where more rigorous statistical interpretations were made. If only a single value (or very few values) are available from Step A, then a comparison with the data in Table 4 will be helpful.

Table 2. *Lead contents of igneous rocks.*

Rock type	Average lead content (gg^{-1})
Gabbro	1.9
Andesite	8.3
Granite	22.7

Table 3. *Typical lead contents of sedimentary rocks.*

Rock type	Proportion all sediments %	Typical lead content gg^{-1}
Shales	75	23
Sandstones	15	10
Limestones	10	7

Table 4. *Recent estimations of the lead concentrations of soil.*

Country	Median [1] or geometric mean[2] gg^{-1} soil	Range	Author
England/Wales	42^1	5-1200	Archer (1980)
England/Wales	42^2	15-106	Davies (1983)
Scotland*	13^2		Reeves & Berrow(1984)
**	30^2		ditto
Missouri[+]	60^2		Davies & Wixson (1985)
++	50^2		ditto

* inorganic soils; ** organic soils
[+] all soils;   [++] estimated normal background

Statistical Interpretation of Soil Data

Step A may provide enough values (20) to make statistical evaluation a necessity and steps F and L certainly will. Statistical methods are used both to summarise and evaluate the data. The simplest descriptive statistic is the mean, the sum of the measurements divided by the number of measurements. Besides calculating the mean, most computer packages (including business software such as spreadsheets) usually provide the standard deviation, a measure of the spread of the values around the mean. It is also necessary to establish the minimum and maximum values. Data assessment should not stop at this point since these parameters do not fully evaluate the data. It is also important that the median value be calculated. This is the middle value when metal concentrations are arranged in order of increasing concentration.

Table 5 illustrates soil lead concentrations from a typical survey (Davies and Ballinger, 1990). The arithmetic data are characterised by a feature which is

Table 5. *Summary data for lead in 174 samples of surface soil in Somerset, England (Davies and Ballinger, 1990).*

	Lead gg^{-1} soil
Arithmetic values	
Mean	183
Median	52
Standard Deviation	798
Minimum	8.0
Maximum	(1.0%)
Log10 transformed values	
Mean	66
Standard Deviation	3.0
95% probability range	
Low	7.6
High	577

common in this kind of data, namely that the mean is greater or very much greater than the median. The most common inference drawn from the value of the mean is one of typicality, the average value in its colloquial sense. But the median is also a measure of central tendency. The two statistics are seen to differ in Table 5 the mean being 3.5 times greater than the median. The distribution is positively skewed and in this instance the median far better represents central tendency than does the mean. Many statistical packages will also provide the skewness or third moment statistic. A positive value indicates a clustering of samples to the left of the mean.

The most commonly used statistical modelling techniques (analysis of variance, regression analysis or correlation analysis) are 'parametric' tests: they require the test populations to be *normally* distributed, *i.e.*, they should not be skewed. Populations can be normalised by transforming the data and a frequently used transformation is to convert each value to its logarithm (the common $\log_{10}$ or the natural $\log_e$). Table 5 shows the result of a $\log_{10}$-transformation. Recalculation and anti-logging yields the geometric mean (66 µg g^{-1}) which is only 1.3 times the median (52 µg Pb g^{-1}). The reduction in spread is reflected in the geometric deviation (3.0) compared with the standard deviation (798). As a general rule, all soil lead data should be log-transformed before statistical analysis.

The way in which the range is quoted needs careful consideration. Of course the observed range should be published as in Table 5. But only the very occasional sample in the study area approaches the observed maximum (1.0% Pb, rounded). A different measure of range must be used if typicality is to be inferred. A property of the normal curve is that the proportion of the area underneath it is described by the mean +/- some multiple of the standard deviation. The mean +/- 2s accounts for 95.5% of the area and the mean +/- 1.96s accounts for 95% of the area. Similarly, mean +/-3s accounts for 99.7% of the area and mean +/- 2.58s accounts for 99% of the area. From this it is useful to quote the 95% probability range (mean +/-1.96s) using the log-transformed data. This was done for the data summarised in Table 5 and indicated that for the Mendip Hills of north Somerset most soil lead concentrations lie between 8 and 577 mg Pbkg^{-1}.

The approach outlined above is not, of course, the only way of summarising voluminous data. But the suggested statistics are easily calculated using a microcomputer and relatively inexpensive software is widely available, especially for the IBM PC and its clones.

Identification of Contaminated Soils

There is no simple, unequivocal way of recognising when a soil has been contaminated by lead since the element occurs naturally in all soils, albeit at low concentrations. The problem of recognising whether contamination has taken place becomes one of deciding whether the measured concentration is within the range of what could occur naturally for that soil or whether the measured concentration is *anomalous*.

Quantitative approaches to the description and evaluation of lead and other trace element data for soils are still in their infancy and it is not clear what is the best model to describe the variability of soil metal concentrations. Ahrens (1954, 1966) has proposed that the distribution of elements in igneous rocks approximates to a log-normal distribution. This model does not necessarily apply to soils but the available evidence suggests it may. Rose *et al.* (1979) discuss the concept of threshold, the upper limit of normal background fluctuations. Values above background are considered anomalous. This approach is directly applicable to contamination studies since a contaminated soil is an anomalous soil. The simplest way of identifying threshold concentrations is by collecting samples from apparently uncontaminated areas (*e.g.*, those remote from urban or industrial influences). After analysis the geometric means and deviations are calculated. The threshold is then the value lying two or more standard deviations from the mean, depending on the probability level required. An anomalous value is one which lies above the threshold. Where more than one sample is apparently anomalous then the differences between the two groups (control and anomalous) can be assessed by standard statistical tests such as the familiar *t*-test.

Very often it is not possible *a priori* to separate contaminated and uncontaminated soils at the time of sampling. The best that can be done in this situation is to assume the data comprise several overlapping log-normal populations. A plot of %cumulative frequency *versus* concentration (either arithmetic or log-transformed values) on probability paper produces a straight line for a normal or log-normal population. Overlapping populations plot as intersecting lines. These are called broken line plots and Tennant and White (1959) and Sinclair (1974) have explained how these composite curves may be partitioned so as to separate out the background population and then estimate its mean and standard deviation. Davies (1983) has applied the technique to soils in England and Wales and thereby estimated the upper limits for lead content in uncontaminated soils. A degree of subjectivity is involved in the interpretation and often plots are not readily partitioned.

It should not be assumed that anomalous concentrations necessarily indicate contamination. Bolviken and Lag (1977) have described areas in Norway where the absence of vegetation is due to the toxic effects of high concentrations of metals in soils as a result of weathering of sulphide ores close to the surface. This is a natural process having nothing to do with contamination.

Identification of a geochemical anomaly should, in the first instance, be considered as only that, an anomaly. Other evidence must be taken into account to decide whether the anomaly is natural or is a neoanomaly, one caused by anthropogenic contamination.

Cartographical Representation of Data

Many ways are possible for representing the spatial distribution of lead data, ranging from sized or coloured symbols, based on the relative concentration at the sample locality, to complex statistical surfaces such as trend surface plots. But whatever style of representation is chosen an essential step in the data reduction is the manner in which the concentration values are classified to produce a relatively few groupings of the data from the minimum to the maximum. This can be done quite empirically by allocating class limits from experience. But this approach involves too high a degree of subjectivity.

The simplest systematic approach is to divide the range by, say, 10. Each metal value may then be allocated to its relevant class and mapped. But skewed data again present problems. For the data in Table 5 the range is approximately 10,000 giving a class interval of 1,000. But only 4 samples contain 1,000 g Pb g^{-1} soil. here again, a log transformation improves matters. For the same data the class interval is (log) 0.3: the lowest class contains 1 value as does the highest and the data are regularly distributed through the classes.

A more laborious but more informative approach is through the frequency distribution of the data. The log-transformed values are classified (a class width of 0.1 is often suitable) and the percentage frequency in each class is calculated. These are then summed to 100%. A plot of concentration versus cumulative percent frequency is constructed and a smooth, sigmoid curve is interpolated between the points. This curve is then used to estimate the concentrations corresponding to selected percentiles. For contamination studies it is often convenient to use the 50, 60, 70, 80, 90 and 95th. percentiles. Ideally, the 50th percentile, the median and the geometric mean should be the same but irregularities in the frequencies combined with a best-fit of the curve often produce small discrepancies. Here, the 50th percentile corresponds to 40 μg Pb g^{-1} soil compared with the geometric mean of 66 μg g^{-1} and a median of 52 μg g^{-1}. The 95th. percentile equates to 450 mg Pb kg^{-1}, whereas the 95% upper probability limit was quoted above as 577 μg g^{-1} (the 97th percentile).

Broadly, there are two kinds of map. Where it cannot be assumed that there is any progressive change across a given area for the value of the parameter under investigation choropleth maps are constructed. Areas of equal value are separated by boundaries from adjacent areas of different values. Familiar examples are soil or geology maps. But where progressive change can be assumed then isoline maps are used. Examples are topographical maps where contours connect points of equal elevation or weather maps where isobars connect points of equal atmospheric pressure. The familiarity of topographical maps compared with other isoline maps has often led to all isoline maps being loosely described as 'contour'.

It is dubious whether all geochemical data are properly representable by isoline maps. Since chemical composition depends on rock type and rock type can be depicted properly only by choropleth maps then isoline maps could be presumed not to be generally suitable for geochemical data. But although soil composition is strongly influenced by parent material composition other processes are also significant, such as wind or water transportation of particles and compounds.

Contaminating sources are generally classed as point or line. A smelter stack is a typical point source and highways are typical line sources due to the movement of motor vehicles and their exhaust emissions along them. A cluster of point sources forms an area source. But whatever the geometry of the source as contaminants are carried away they become diluted. Fallout from a stack tends to decline exponentially away from the source. Overbank inundation in river systems leads to greatest contamination nearest to the river channel. Distinctive depositional patterns are thereby created and much can be inferred about the presence and nature of contamination by studying these patterns. It is reasonable therefore to conclude that isoline maps are often suitable for the study of lead contamination.

A number of computer program packages are now available for constructing isoline maps. There are major mainframe computer packages which produce very high quality monochrome or colour plots with inkjet or thermal printers. There are also some good packages available for use with desktop computers.

Whichever system is used there is an important first stage. The data are imported into the program as X, Y and Z values (two geographic coordinates and the lead concentration) and from these a uniform grid of values is created. This entails extrapolation between neighbouring values to calculate the concentration at the grid intersection. The most common involves searching over a defined radius around each sample point and averaging using a weighting factor dependent on the inverse square of the distance between points. Since production of a regular grid is an essential preliminary then the more the distribution of the original data departs from regularity the more possibility there is of distortion of the eventual geographic pattern and the higher the likelihood of misinterpreting the pattern. Where the terrain permits it is much better to sample on a grid basis rather than rely on the chosen computer algorithm to regularise the grid.

Contaminated Soils

The lead content of the main rock types is not so variable as to suggest there are large differences in concentrations in uncontaminated soils of different parentage. The main interest in elevated soil lead values therefore centres on increases due to human activity. In general, lead becomes environmentally labile whenever the metal or its compounds are heated, dissolved or pulverised. Soil is a major sink for anthropogenic lead and there are several well recognised major sources, namely, mining and smelting, sewage sludge usage in agriculture, and present and past contamination from vehicle exhausts. There are also other local sources, *e.g.* lead arsenate ($PbHAsO_4$) has been applied to orchard trees to control insect pests and orchard soils may therefore contain elevated concentrations of Pb (Frank *et al.*, 1976; Merry *et al.*, 1983). Commercial use of these sprays is now infrequent since they have been replaced by organic pesticides.

It is of interest that soil concentrations in the USA appear to be lower than those found in Britain. Pierce *et al.* (1982) determined metal concentrations in 16 soil series in Minnesota, USA, in order to establish baseline levels. Total lead was below 25 μg g^{-1} for all the soils. Holmngren *et al.* (1993) have reported values for 3045 surface soil (Ap horizon) samples from cropland in the USA. The median concentration was 11 μg Pb g^{-1} and the mean was 18 μg Pb g^{-1}. The relatively high values observed in British topsoils may indicate a widespread low level contamination arising from more than two centuries of industrial and metallurgical activity. This hypothesis is supported by higher values found during studies in south eastern Missouri, USA, a long established lead, copper, nickel and cobalt mining area where the first lead mine was opened in the late eighteenth century (Davies and Wixson, 1985).

Lead Derived from Vehicle Exhausts

From 1923 onwards petrol contained lead alkyls as anti-pinking or anti-knock additives. After 1945 car ownership became widespread and it is now recognised that lead in vehicle exhausts is a major environmental contaminant. Figure 2 shows lead usage in paint and petrol from 1910 to 1989. The rapid decline in the use of lead alkyls in petrol after 1972 is very clear.

Warren and Delavault (1960) were the first workers to discover that lead was accumulating in the roadside ecosystem. Having provided new information relating soil contents to bedrock they remarked that the contributions of lead from petrol fumes and orchard sprays (lead arsenate) merited attention. This paper was soon followed by a report by Cannon and Bowles (1962) which demonstrated that grass was contaminated by lead within 500 feet downwind of highways in

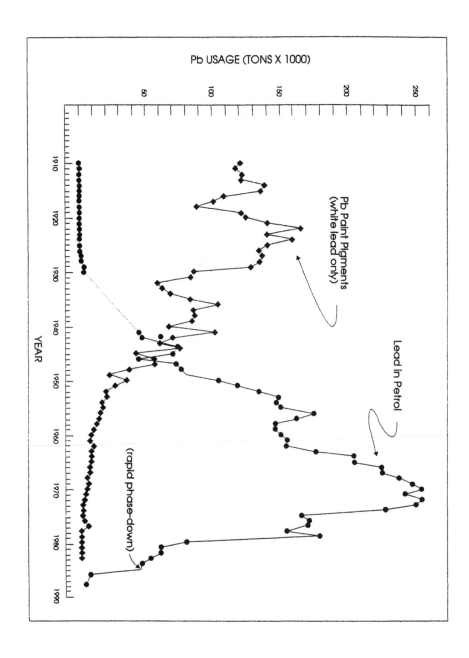

Figure 2 *Lead usage from 1910-1989 (U.S. Bureau of Minesota).*

Table 6. *Lead contents of soil at three distances from typical main roads, calculated from data in Smith (1976).*

Distance from road (m)	Geometric mean probability range	95%	Number of sample
	$\mu gPb\ g^{-1}$ soil		
<10	192	18–2,017	20
15	161	50–511	6
>30	53	14–203	17

Denver, Colorado and the relationship between lead content and distance from the road was exponential. Contamination of the roadside environment has now been reported by many authors in many countries and it must be counted a world wide consequence of the use of leaded petrol.

The concentration of lead in roadside soils depends on traffic densities, topography, climate and wind conditions. Most authors agree that the distance-decline relationship away from the road is curvilinear and local background levels are reached by 30 to 50 m from the road. Wheeler and Rolfe (1979) modelled the deposition of lead by a double exponential function of the following form:

$$Pb = A_1^{-ekD} + A_2^{ek.D}$$

A_1 and A_2 were linear functions of average daily traffic volumes and the two exponents (e and constants k) were assumed to represent two families of particles of different sizes. Larger particles were deposited within 5 m of the road and the small particles within 100 m.

Data in a review by Smith (1976) have been recalculated to provide typical roadside lead concentrations (Table 6). The values in Table 6 demonstrate that lead contamination does not extend appreciably beyond 30 m from the road. Within 10 m the mean soil concentration is 4.6 times the average level (42 g g^{-1}) reported by Davies (1983) for uncontaminated British soils. This is consistent with Smith's own assessment that if the background soil lead level is taken as 20 $\mu g\ g^{-1}$ then at 10 m distance from the highway the lead level is typically only 5-15 times the background.

Lead in Paint

From the late eighteenth century until the late nineteenth century the only white pigments used in paint were lead compounds. Since the end of the nineteenth century paint formulations have included TiO_2, $BaSO_4$, $CaSO_4$ and $CaCO_3$. White paints have generally relied on three carbonates of lead, namely, normal lead carbonate, $PbCO_3$, and the basic lead carbonates, $2PbCO_3.Pb(OH)_2$ and

$4PbCO_3.2Pb(OH)_2.PbO$. Other white lead compounds in use include $PbSO_4$ and the silicate $3PbO.2SiO_2.H_2O$. Red colours can be based on red lead (Pb_3O_4) and the chromates ($PbCrO_4$ and $PbCrO_4.PbO$) provide the yellow colours for street markings. Since 1977, paint intended for use on exposed interior and exterior residential surfaces in the U.S.A. should not exceed 0.06% Pb. The Paintmakers Association of Great Britain agreed that decorative paint in Britain would be free of added lead by 1987 and would then comply with the U.S.A. limit.

Lead in Urban Soils

Towns and cities are foci for traffic and the effects of roadside contamination are relatively greater in urban areas. But lead is not the only contaminant of urban soils and petrol is not the only source of lead. Purves (1966) appears to have been the first worker to have recognised elevated levels of trace elements in urban soils. A study of garden soils in Edinburgh and Dundee, Scotland, indicated substantial contamination by copper and boron. The author speculated that soot and domestic chimney smoke might make a substantial addition of these elements to garden soils.

In London, England, Davies *et al.* (1979) reported a range of 42 – 1,840 µg Pb g^{-1} soil (EDTA-extractable). The values could be zoned according to the proximity to the city centre: means for three zones were 523 µg g^{-1} in the 0–4 km region, 242 µg g^{-1} in the 4–10 region and 142 µg g^{-1} in the 10–30 region. There was an approximately exponential decline of soil lead content with increasing distance from the centre. A similar curvilinear relationship was found (Davies and Houghton, 1984) in the Birmingham area where soil lead and distance from the city centre were related by the polynomial regression

$$Y = 272.5 - 27.5X + 0.82X^2 \quad R^2 = 0.528$$

where Y = distance in km and X = soil lead content.

A recent survey of 100 households in each of 53 representative towns and city boroughs in Britain confirmed elevated concentrations of lead in surface garden soils compared with agricultural soils (Culbard *et al.* 1988). The gardens sampled were selected on a random basis within each town; the towns were chosen on the basis of population, geographical location and degree of industrial/urban development. Lead concentration ranged from 12 to 10^4 µg g^{-1} (geometric mean 230 µg g^{-1}) in 3,550 garden soils from locations excluding London and areas affected by mining and smelting. In the 579 gardens sampled in 7 London Boroughs the mean value was 647 µg g^{-1} Pb. Agricultural soils in Britain usually contain from 10-150 µg g^{-1} Pb. Lower lead values in soils from households of less than 30 years of age compared with those of 100 or more years of age were considered by the authors to reflect (a) voluntary reduction in lead content of household paints and (b) a shorter history of soil contamination from bonfires and fossil fuel residues from open coal fires and soot commonly used as soil improvers and ameliorants.

Lead contamination is not confined to the gardens of major towns or cities. Davies (1978) reported the lead content (EDTA-extractable) of soils from 87 British gardens in both rural and urban environments: 63% of the soils were contaminated when compared with rural arable norms. Although the highest values were derived from lead mining areas and city gardens, older rural gardens were also mildly contaminated. In two villages in Devon and Cornwall significant correlations were found between the age of the house or garden (i.e., duration of occupancy) and the soil lead content.

Studies of lead in soil have demonstrated a greater contrast in North American towns and cities than are described for Britain. A study of urban vegetable gardens conducted by Mielke (Mielke *et al.*, 1983) in Baltimore, Maryland revealed a decisive clustering (probability (*p*) <10-23) as predicted by Davies and Houghton (1984) of high lead soils in the inner-city. The geographic patterns of soil lead in cities requires the integration of multiple sources of lead with the characteristics which influence each of the sources. For example, the two major lead sources, lead-based paint and lead in petrol, are influenced differently by city size and age of city. The historic use of petrol varies as a function of traffic volume and congestion and increases exponentially with city size. This means that larger cities should have significantly more lead in their soils than small towns.   The historic use of exterior lead-based paint varies with age and construction qualities (*e.g.*, type of siding and lead content of paint) of housing found in communities of a given city. It is expected that lead in soil near the foundations would be approximately similar for the same age and quality of housing. Comparison of non-mining and non-smelting cities of Minnesota and Louisiana by Mielke (1993) revealed that the amount of lead in soil is related to urban location and especially city size and that age of construction does not explain the amount of lead in soil within a given community. For example, Natchitoches is the oldest town in Louisiana. However, the median lead content of soil collected within 1 metre of foundations in Natchitoches is 33 μg g^{-1} (range 6-1430) compared with 840 μg g^{-1} (range 8 – 69,000) for foundation soils of Central New Orleans (p-value 10-21). The same pattern has been described between the inner-cities of the larger cities of St. Paul and Minneapolis compared with the small town of Rochester, Minnesota (Mielke *et al.*, 1984; Mielke, 1993). It is noted that foundation soils generally contain several times more lead than soils near roadways in inner areas of large cities and that foundation soils contain smaller amounts of lead than roadside soils in suburbs and small cities (Mielke *et al.*, 1988; Mielke and Adams, 1989; Mielke, 1993). See Table 7. As noted above, the shorter history and less dense industrial development may account for the lower lead level in rural soils of the US. Also, the greater reliance on the use of cars in US cities may account for the higher constrast of lead in urban soil of the US cities compared with British cities.

Contamination from Mining and Smelting

The first report of the influence of lead mining on the health of crops and stock

Table 7. *Lead contents of city soils in the USA (Mielke et al., 1983 and Mielke, 1993).*

USA city and sample type	N	Median	Range g g^{-1}
Baltimore gardens	422	100	1 – 10,900
Louisiana			
Central New Orleans			
Foundation	201	840	8–69,000
Streetside	723	342	4–950
Open space	74	212	10–10,600
Outer New Orleans			
Foundation	220	110	1–24,400
Streetside	765	110	1–6,340
Open space	80	40	2–3,960
Suburban New Orleans			
Foundation	332	50	2–5,650
Streetside	1192	86	2–2,150
Open space	114	28	4–540
Natchitoches			
Foundation	18	33	6–1,430
Streetside	66	50	10–550
Open space	6	14	8–36
Minnesota			
Minneapolis			
Streetside	168	265	
Mid yard	265	290	
Foundation	184	860	
St Paul			
Streetside	169	160	Mielke *et al.*,1989,
Mid yard	200	205	Mielke 1992.
Foundation	99	675	
Rochester, MN			
Streetside	27	58	
Mid yard	22	30	
Foundation	18	50	

was written by Griffith (1919). During the years 1908–1913 he investigated frequent complaints of soil infertility from farmers in the Aberystwyth area of Wales (Great Britain). He concluded that the primary cause in most cases was the presence in soil of appreciable quantities of lead which had originated from lead mining in the previous century.

Griffith's early report was not followed up for fifty years. Then, Alloway and Davies (1971) described a comprehensive investigation of lead the Aberystwyth district. They confirmed the work of Griffith and found that alluvial soils in the Ystwyth river valley contained 90–2,900 (mean = 1,419) μg Pb g^{-1} soil compared with 24–56 (mean = 42) μg Pb g^{-1} soil in a neighbouring valley which was unaffected by mining.

In north-east Wales (Davies and Roberts, 1975) agricultural soils contain 146–4,795 (mean = 3,470) μg Pb g^{-1} and garden soils contain 920-5,897 (mean = 1,604) μg Pb g^{-1} . These authors published computer drawn Pb isoline maps for the area and they calculated that there are 171 km^2 of land containing 100 μg Pbg^{-1} soil and 47 km^2 have soils with lead contents ranging from 1,000-10,000 μg Pb g^{-1} (Davies and Roberts, 1978). A similar approach was used in a baseline survey in Missouri, USA, by Davies and Wixson (1985). The survey was conducted in Madison county where mining for lead first began in the late eighteenth century and in the following 150 years there was extensive mining especially for ores of copper and nickel. The maximum soil lead concentration was found to be 2,200 μg g^{-1} but 95% of the samples contained <355μg Pb g^{-1}. Isopleth plots of the soil lead contents showed a strong association between concentrations >355 μg Pb g^{-1} soil and old mine workings, dressing floors and railway ore loading places. Thus, contamination of agricultural land by lead in old lead mining areas is commonplace.

Surveys around smelters have shown maximum accumulations close to the stack and a rapid decline away. The distance-decline curve is often exponential. Lagerwerff et al. (1973) worked in the area of a smelter at Galena, Kansas where production started in 1903. At 330 m to the northeast of the plant the top soil contained 1,600 μg Pb g^{-1} compared with 428 μg Pb g^{-1} at 1,670 m to the northeast. At Kellogg (Idaho) smelting started at the end of the nineteenth century (Ragaini et al., 1977). Accumulations of lead in the top 2 cm of soil were explained by deposition of fallout and levels varied according to wind direction: within 2 miles of the stack the authors reported 7,600, 6,700, 5,300 and 1,700 μg Pb g^{-1} soil.The modern mining and smelting industry is comprehensively regulated in most countries so that emissions of lead to the environment are minimised. But the large scale of the industry inevitably leads to accumulations of lead in soils around the works. In addition, some of the modern mines and smelters are located on the sites of older, dirtier works which have left a legacy of environmental contamination.

Sewage Sludges and other Organic Residues

A source of soil lead which has achieved considerable prominence in recent years is sewage sludge. Farm yard (barn yard) manure has long been recognised as a desirable soil amendment material. It is a useful short and long term source of the essential plant nutrients nitrogen, phosphorus and potassium as well as a source of essential plant and animal micronutrients. But supplies of manure are often limited in areas of arable cropping and treated, pathogen-free sewage sludge is an economical alternative. Unfortunately, sewage sludge can have one undesirable property, namely, a high content of heavy metals. Coker and Matthews (1983) state that the typical concentration of lead in human faeces is 11 μg g^{-1}. Additional lead is derived from the use of lead water pipes and road run-off. Furthermore, it is common practice for industrial effluents to be passed into the foul (sanitary) sewer and organic residues effectively remove metal ions by complexation. Thus the metal contaminants are retained in the sewage sludge. The agronomic implications of this were not realised until Le Riche (1968) reported metal contamination of soil following the long term application of high rates of sewage sludge on an experimental market garden. Since that date there has been a proliferation of reports concerning metals in sludges and in crop plants grown on sludge amended soils. Several countries now have advisory limits to regulate the use of the material on farmland. Most attention has probably been paid to cadmium but lead has also been the subject of much research.

There are several published reports of metal concentrations in sewage sludges. Berrow and Webber (1972) analysed 42 sludges from rural and industrial towns in England and Wales. The lead contents of the dried materials ranged from 120–3,000 μg g^{-1} (mean = 820 and median = 700 μg Pb g^{-1} respectively). Sommers *et al.* (1976) reported a range of 545–7,431 μg Pb g^{-1} for sludges from 8 Indiana (USA) cities.

The additions of lead to soil from sewage sludge are different from accumulations from urban and industrial sources in two significant ways. First, lead in sludge is incorporated into soil in an organic form rather than an inorganic form. It is not clear what the implications for plant availability are. Second, it is not possible to predict which fields have elevated metal levels since usage of sludge is a management decision by an individual farmer. This contrasts with lead contamination from other sources where the presence of a smelter, mine, road or urban centre is in itself enough to suggest the possibility of some localised accumulation of lead in soils.

VI. BIOAVAILABILITY OF LEAD FOR ANIMALS

The bioavailable fraction of the total quantity of lead in a diet is generally defined as that fraction which can be absorbed into the blood stream by the animal species ingesting the diet. Because food constituents modify lead absorption, the bioavailability of lead in test materials such as soil is compared with that of the water soluble lead salt, lead-acetate, which is considered 100% bioavailable. Research has shown that many factors can influence whether lead in soil and dust ingested by children is actually absorbed into the blood. The physical and chemical properties of the dust particles, the nutritional status of the children, and whether the soil is ingested with food or between meals, can each substantially affect whether soil lead is absorbed. These factors can so strongly affect lead absorption that decisions to reclaim or even replace contaminated soils should take into account the chemical nature of the lead and the organic content and pH of the soil involved.

Another very important consideration is the relationship between increasing soil ingestion and lead absorption. If a child exhibiting *pica* is to be protected, soil ingestion by a child in the 95th or 99th percentile of the ingestion range (0.5 and 5.0 g/day, respectively; Calabrese *et al.*, 1989) must be considered. However, if lead absorption approaches a plateau with increasing soil dose, the same child may be at no greater risk than the median soil-ingesting child. The adsorption (*i.e.*, recapture) of lead on to soil particles in the small intestine would be expected to cause this response pattern. Similarly, the adsorption of lead by soil particles in the intestine is hypothesised to allow low soil lead concentrations to be less available compared with higher soil lead concentrations when soil adsorption plays a less significant role. The response to increasing soil ingestion has been found to level off in several studies. Approximately 300 µg Pb g^{-1} was found to be a "no effect" level for lead in sewage sludge compost ingested by cattle (Chaney *et al.*, 1989). Below this level lead in tissue did not increase even though the cattle ingested significantly increased amounts of total lead. Thus, the risk from soil lead is very different than the risk from water-lead, paint-lead, or food-lead because these lead sources do not provide a lead readsorption capacity.

A. Factors That Influence Risk of Soil Lead

Lead concentration in soil, the size of soil particles rich in lead, the nature of the chemical species of lead found in soil and nutritional factors together with human behavioural traits interact in controlling the risk from soil lead (Mielke and Heneghan, 1991). It is clear from many studies that children vary remarkably in their blood lead (PbB) level when exposed to similar lead sources. Parental supervision, personal habits (mouthing of fingers, hands, toys; chewing fingernails; washing hands), pica and the quality of nutrition vary so greatly among children that some children may be at a relatively low risk even when they live or play in areas with high soil lead. However, for risk assessment one must consider either the child described by Duggan and Inskip (1985), the average child playing in a normal dirty way, or, the "most-exposed, most-susceptible individual" which is the US-EPA approach. The most exposed individual is a poorly-supervised child who is regularly exposed to lead-rich soil, exhibits pica and whose nutrition is poor in relation to factors which interact with lead absorption. Such a child would therefore ingest more soil lead and absorb a higher percentage of this soil lead than would the well cared for and well nourished child living in a well kept environment.

B. Bioavailability of Ingested Soluble Lead

Research on laboratory animals over many years has characterised the effect of nutritional factors on lead absorption (Mahaffey, 1981, 1985; Mahaffey and Michaelson, 1980). More recently, adult human lead isotope absorption studies and lead balance studies in infants have clarified the understanding of human lead absorption. In addition, several feeding studies using livestock and laboratory animals have directly tested lead absorption from dietary soil/dust.

C. Effect of Lead Compound and Particle Size on Lead Absorption

Research has been conducted to evaluate the bioavailability of lead in different lead-compounds. Allcroft (1950) reported on long-term feeding studies in which several lead compounds were fed to cattle and reported great differences between both lead compounds and different particle sizes of the same compounds. In particular, large particles of PbS were much less toxic and caused lower tissue levels of lead than did small particles of PbS or other compounds.

Barltrop and Meek (1975, 1979) studied the bioavailability for rats of different lead compounds and paint pigments (of various particle sizes) using a 48 hour feeding period. In their work, larger particles of several materials were of lower bioavailability than smaller particles. This would appear to result from the poorer dissolution of larger particles during the short exposure period in the acidic environment of the stomach. Compounds which are readily soluble in weak acids were highly bioavailable. Thus, some lead pigments, lead ores and paint particles are only partially dissolved in the stomach. Human feeding studies of Rabinowitz *et al.* (1980) tested the absorption of finely divided PbS by fasting human volunteers. In this test, PbS was highly bioavailable, but this particular PbS preparation did not simulate lead ore or lead ore wastes. Healy *et al.* (1982) tested the solubility of different particle size preparations of PbS in gastric fluid. This work confirmed that smaller particles of PbS were dissolved more rapidly than larger particles. Their work focused on the bioavailability of PbS from cosmetics (*e.g.* surma) which appear to be transferred to the mouth after hand contact (Healy *et al.*, 1982; Healy, 1984). The implication of these findings for soil or dust lead is that larger particles of PbS (*e.g.* galena ore particles dispersed by mining, transport, or smelting) would be expected to have a significantly lower bioavailability than other sources of lead in soil.

D. Effect of Nutritional Factors on Lead Absorption

Feeding studies to assess the effect of nutritional status on lead absorption have shown that a deficiency of calcium or iron increases lead absorption (Mahaffey, 1981, 1985). When dietary intake of calcium fell below about 50% of the dietary levels recommended by the National Research Council (NRC) for growing rats, lead absorption increased substantially (Mahaffey *et al.*, 1977). Mahaffey (1981) has summarised these nutritional interactions in relation to known dietary limitations in urban poor children and pregnant women, the largest groups who comprise the "most-susceptible" individuals for excessive soil lead.

Iron deficiency was found strongly to affect lead absorption in rats. lead absorption declined with a further increase in dietary iron above the minimum dietary requirement. Because many children are iron deficient, this nutrient could be important in assessing risk of soil lead ingestion. According to Angle *et al.* (1975), results from two independent research programmes found different effects of Fe-deficiency on lead absorption by adult humans. A report by Watson *et al.* (1980) showed that individuals with low serum ferritin (indicating low body iron reserves), who absorb an increased fraction of dietary Fe, also absorbed increased amounts of carrier free lead. However, Flanagan *et al.* (1982), using [203]Pb with 200g of carrier lead, found no effect of iron status (serum ferritin) or added dietary iron on lead absorption by humans. This was a direct test of the Watson *et al.*(1980) report, but it used an improved experimental design. More study is needed on the influence of iron.

One of the most important findings of this human lead absorption research was that lead absorption is greatly reduced by simultaneous ingestion of food (Blake *et al.*, 1983; Flanagan *et al.*, 1982; Heard *et al.*, 1983; James *et al.*, 1985; Chamberlain, 1987; Rabinowitz *et al.*, 1980) compared with lead ingested during fasting. The effect of a meal on lead absorption lasted about 2-3 hours after eating because of slow gastric emptying after a meal (James *et al.*, 1985). This result has important implications for the absorption of soil/dust lead compared with water lead or paint lead when ingested between meals. Studies concerning dietary components which reduce lead absorption include calcium phosphates and fibre (Blake *et al.*, 1983; Blake and Mann, 1983; James *et al.*, 1985). Combinations of calcium and phosphorus had more effect on lead absorption than did calcium alone (Blake and Mann, 1983; Heard *et al.*, 1983). In other work, lead isotopes were incorporated into lamb liver and kidney, and into spinach to allow comparison of lead intrinsic to a food with a lead isotope extrinsically added to a meal with that food. This research showed that "food" lead was absorbed in amounts equal to lead salts added to an equivalent meal (Heard *et al.*, 1983).

Lead in oysters was about 70% bioavailable to Japanese quail as lead acetate added to the purified diet (Stone *et al.*, 1981). James *et al.* (1985) evaluated a number of meals and dietary components. Test meal components such as phytate or EDTA (ethylenediaminetetraacetate) reduced lead absorption compared with the effect of an equal amount of calcium and phosphorus in a low phytate, refined diet, basal meal (James *et al.*, 1985; Flanagan *et al.*, 1982). On the other hand, milk in a meal increased lead absorption compared with the expected effect of an equivalent amount of calcium and phosphorus in the milk. Thus, phytate, fibre and calcium in whole grain foods would tend appreciably to reduce lead absorption compared with more highly refined grain products.

As observed with other nutrients, lead absorption is proportional to the activity of free Pb^{2+} ions in the intestine. For example, addition of EDTA to a test diet reduced absorption (Flanagan *et al.*, 1982; James *et al.*, 1985) because chelation of an element reduces its chemical activity. The role of phytate (James *et al.*, 1985; Wise, 1981), and some tannin and fibre components (Peaslee and Einhellig, 1977; Paskins-Hurlburt *et al.*, 1977) should be similar to EDTA. Soil and dust should act like fibre in this regard, by adsorbing lead and reducing Pb^{2+} ion chemical activity, thereby reducing soil lead bioavailability.

These results on the relationship between calcium and phosphorus concentration in human diets and lead absorption raise questions concerning how calcium and phosphorus interfere with lead absorption. There are several possible mechanisms. One is a simple interaction of calcium and lead at the intestinal lead absorption site. The most likely mechanism for the effect of calcium and phosphorus on lead absorption appears to be co-precipitation of lead with calcium-phosphates formed during the digestion process. Co-precipitation of Zn with calcium-phosphates indicates how this might occur. Nelson *et al.* (1985) first studied model systems and found that calcium-phosphate formed quickly (within minutes) when calcium and PO_4 were present at levels common in whey, the solution remaining after curds form in acidified cow's milk. Further study showed that Zn coprecipitated with the calcium-phosphates above pH 5.0 as the pH increased and co-precipitation was complete by pH 6.0 (Nelson *et al.*, 1987).

Co-precipitation may be the model for the effect of dietary calcium and phosphorus on lead absorption. The clear interaction of dietary calcium and phosphorus concentrations in animal and human studies of lead absorption conform with a solubility product model.Lead co-precipitation with calcium-phosphate should be complete at a pH even lower than that found for Zn.

E. Bioavailability of Lead in Ingested Soil and Dust

One approach to the problem of the bioavailability of lead in soil is to ask what happens to wildlife which live in areas with high soil lead. Wildlife are unable to avoid exposure to and usually show at least some absorption from soil/dust lead in their habitat (Elfving *et al.*, 1978; Hutton and Goodman, 1980; Ireland, 1977; Scanlon *et al.*, 1983; Young *et al.*, 1986). These studies suggested appreciable

Table 8. *Effect of the daily dose of igested dust-Pb on Pb in tissues of rats fed Queens, NY tunnel dust (sieved) mixed in purified diet for 42 days before analysis of tissues. Dust contained 2.2% Pb/kg (Stara et al., 1973).*

Daily	Blood	Tissue Pb			
Dose mg Pb d^{-1}	Peak Pb g dl^{-1}	Femur	Kidney	Liver	Brain
		----g Pb g^{-1} tissue ----			
0.0	11	0.75	-	0.13	0.032
0.5	33	25.7	2.5	0.56	0.030
1.0	39	33.6	3.2	0.52	0.055
5.5	51	66.1	9.4	1.3	0.28

bioavailability of soil/dust lead, but did not make specific comparison with soluble lead salts added to control diets. Similarly, livestock grazing pastures on soils rich in lead (mine wastes, naturally lead-rich soils) had increased lead levels in body tissues indicating that soil lead was at least somewhat bioavailable, but much less available than soluble lead sources (Allcroft, 1950; Egan and O'Cuill, 1970; Harbourne *et al.*, 1968; Wardrope and Graham, 1982).

The contribution from lead in mining wastes to blood lead has recently been considered in a comprehensive review by Steele *et al.* (1990). An evaluation of studies for mining areas found no strong correlation between soil lead and blood lead and no elevated blood lead concentrations in areas with very high soil lead concentrations (Heyworth *et al.*, 1981). At the lower end of the range the regression gradient was not steep as reported by the EPA (Barltrop *et al.*, 1975). The report notes that while epidemiological studies may not be conclusive, when viewed together, they do indicate that mining wastes may be different from other soil/dust lead sources in contributing to blood leads.

Crystalline PbS is the form of lead remaining in mine tailings after extraction of the ore. The extensive report by Steele *et al.* (1990) indicated a possible reason for a reduced impact on child blood lead of lead sulphide in soils contaminated by mine wastes. Mine wastes have larger particle sizes which decrease the bioavailability of lead in the gastrointestinal tract. Lead sulphide resists absorption in the gastrointestinal tract when compared with other forms of lead.

Several studies were conducted to test directly the bioavailability of soil/dust lead to animals. Stara *et al.* (1973) reported studies of rats fed road tunnel, highway, or smelter dusts. Accumulation of lead in bone or kidney was non-linear with dose (Table 8), with lower %-absorption as dose increased. Bone lead tended to approach a plateau as the amount of soil lead in the diet increased. Another way to view these results is that the highest %-absorption of soil lead occurred at the lowest soil ingestion level. Table 9 shows results from their comparison of the absorption of lead by rats fed different dust sources. El Paso

Table 9. *Effect of dust source and Pb in tissues of rats fed dust supplying 1 mg Pb d⁻¹ for 36 days.* (*Stara et al., 1973*).

Dust Source	Dust-Pb	Blood	Tissue Pb			
	%	Peak-Pb g dl⁻¹	Femur	Kidney	Liver	Brain
				g Pb g⁻¹ tissue		
Control	-	12	0.75	-	0.13	0.032
NY Tunnell	2.22	45	33.6	3.2	0.52	0.055
LA Freeway	1.04	39	32.5	2.7	0.32	0.094
El Paso Smelter	0.67	32	23.6	2.5	0.36	0.035

Table 10. *Bioavailability of soil Pb compared with Pb acetate fed to rats at 50 mg/kg diet for 30 or 90 days mixed in a laboratory chow diet (Dacre and TerHaar, 1977). Roadside soil fed at 2.15% of diet; houseside soil fed at 5.0% of diet.*

Diet	Soil Pb concn. mg kg⁻¹	Measured diet Pb mg kg⁻¹	Bone Pb mg kg⁻¹		Kidney Pb
			@ 30 days	@ 90 days mg Pb kg⁻¹	@ 90 days
Control	-	0.6	1.71 a	1.27 a	0.25 a
Roadside	2300	56.0	5.93 b	4.48 b	0.77 b
Houseside	990	51.8	5.20 b	4.98 b	0.76 b
Pb acetate	-	49.1	5.21 b	6.28 c	0.96 c

Means in column followed by the same letter are not significantly different.

smelter dust (0.67% Pb) had an appreciably lower effect on blood lead than Queens Tunnel (New York) dust (2.22% Pb) or Los Angeles freeway dust (1.04% Pb). These rats were fed 1 mg Pb d⁻¹ as dust in gelatin capsules which is equivalent to about 100 mg Pb kg⁻¹ diet. Bone and kidney Pb were also lower for the lower lead concentration smelter dust than for the tunnel or freeway dusts. These tests did not compare absorption of lead from dusts with lead acetate. This work used a purified diet rather than a lab chow diet which favoured lead absorption.

Dacre and Ter Haar (1977) evaluated the bioavailability of lead in garden soil (990 mg Pb kg⁻¹) and roadside soil (2,300 mg Pb kg⁻¹) compared with lead acetate. Equal lead amounts (50 mg Pb kg⁻¹) were added to each diet. A high fibre, high nutrient rat chow was used as the basal diet. Much lower blood-lead and bone-lead levels were reached in their experiment than was found by Stara *et al.* (1973) (Table 10). Bone and kidney lead concentrations after 90 days of feeding showed that soil lead had significantly lower effects than lead acetate. Bone results indicated that soil lead was about 70% as bioavailable as lead acetate.

Chaney *et al.* (1984) reported data from a more detailed evaluation of the

Table 11. *Effect of soil on bioavailability of Pb to rats, and bioavailability of Pb in urban garden soils.*

Treatment[1]		Pb in Tibia	Pb absorption
PbOAc	soil	mg kg^{-1} tibia ash mean std. err.	compared with that of Pb acetate, %
-	-	0.3 0.3 e^2	-
-	11^3	0.0 e	-
+	-	247.0 10.1 a	100
+	11^3	130.0 29.5 bc	53
-	706	40.0 6.1 de	16
-	995	108.0 26.3 c	44
-	1080	37.1 7.3 de	15
-	1260	53.6 7.4 d	22
-	10240	173.0 21.8 b	70

[1]A purified casein-based complete diet was fed to Fisher rats for 30 days. Pb acetate and garden soils were added to supply 50 mg Pb kg^{-1} dry diet. The experimental garden soils comprised 7.08, 5.02, 4.64, 3.95, and 0.488% of the dry diet, respectively; control soil was fed at 5% of diet.
[2]Means followd by the same letter are not significantly different ($P = 0.05$) according to duncan's Multiple Range Test.
[3]Unpolluted farm soil near Beltsville, MD, (Chillum silt loam, 11 mg Pb kg^{-1}), similar to original soil in the urban gardens used in this experiment.

effect of soil on dietary lead absorption. Rats were fed purified complete casein-sucrose diets with and without 5% uncontaminated soil and with and without 50 mg Pb kg^{-1} diet. The lead was added in the form of lead acetate. (50 mg Pb kg^{-1} diet and 50 g soil kg^{-1} diet = 1,000 mg Pb kg^{-1} "soil" during test). Rats were also fed five Baltimore urban garden soils to compare bioavailability of real lead-rich urban soils with that of lead acetate. All lead was added at 50 mg Pb kg^{-1} dry diet (unequal soil amounts). The results are shown in Table 11. Bone lead concentrations were used to evaluate diet lead bioavailability. The addition of 5% uncontaminated soil to the diet reduced lead acetate control to 53% of lead acetate alone. Four soils with about 1,000 mg Pb kg^{-1} yielded bone lead about 24% (15–44%) that of lead acetate, while a garden soil with 10,240 mg Pb kg^{-1} yielded a bone lead content of 70% that of the lead acetate control. Lead in urban soils was appreciably less bioavailable than was lead-acetate freshly mixed with soil. In contrast to Stara's work the general trend showed increased soil lead bioavailability at the high soil lead concentration; *i.e.*, when soil lead concentrations were higher, percent soil in the diet was correspondingly lower.

Table 12. *Effect of laboratory chow versus purified diet on absorption of Pb from paint chips fed at 1% of diet for 35 days. Paint contained 10% Pb as Pb-octoate; diets contained 1000 mg Pb/kg (Mylroire et al., 1978). Diet type affected tissue Pb concentration in each tissue (P = 0.01).*

Diet type	Pb-blood g dL^{-1}	Pb-femur	Pb-liver g g^{-1} wet weight	Pb-kidney
Lab Chow	<10	97	0.24	5.6
Casein-sucrose	140	400	2.7	300.

This may be expected because soil acts like a fibre and a calcium and phosphorus source in the diet. These properties should allow soil to adsorb lead in the lumen of the intestine and reduce net lead absorption. More recent studies by Davis *et al.* (1990) and van Wijnen *et al.* (1990) suggest that the mean values and extremes of soil ingestion may not be as great as cited above.

The use of purified diets (Chaney *et al.*, 1984) yielded much greater lead absorption from dietary soil than that found by Dacre and Ter Haar (1977). The experimental protocols were similar, and dietary lead was fed at the same level. Lead in tibia ash reached 247 mg Pb kg^{-1} in rats fed 50 mg Pb kg^{-1} (Pb acetate) purified diet, but femur lead reached only 5–7 mg Pb kg^{-1} in rats fed the same level of Pb in a lab chow diet (Dacre and Ter Haar, 1977). Mahaffey and Michaelson (1980) discussed this phenomenon and stated it resulted from the much higher levels of calcium, iron and fibre in chow diets compared with "NRC" purified complete diets. A similar effect of diet was observed by Mylroie *et al.* (1978) in a direct comparison of diet type on the absorption of paint lead (Table 12). Much higher bone and and mineral kidney lead were reached using the purified complete rat diet normally recommended for toxicology studies. In many ways, these purified diets are similar to US diets because of low fibre levels (Mahaffey, 1985).

Research on risk from lead in ingested sewage sludge also provides data relevant to the question of bioavailability of urban soil lead. Sewage sludge has been added to usual (or practical) diets of livestock to evaluate food chain transfer (bioavailability) of lead and other potentially toxic materials in the sludges (Hammond, 1980) Substantial percentages of sludge in diets were used to simulate poor livestock management (worst case) situations in which cattle ingest up to 14% soil (Fries *et al.*, 1982). Sludges used in these studies contained varied levels of lead and other elements such as iron and calcium known to interact with lead. Studies reported by Kienholz *et al.* (1979) and Johnson *et al.* (1981) in which cattle ingested sludge with 780 (Table 13) or 466 mg Pb kg^{-1}, respectively, found increased bone, liver and kidney lead. Response was curvilinear, with the slope of increase in tissue lead decreasing at higher sludge dose.

In other studies, cattle grazing pastures amended with sludges or sludge

Table 13. *Effect of percentage of sewage sludge in diet on Pb residues in tissues of cattle which consumed the test diets for 180 days (Kienholz et al., 1979). The digested sludge was from Denver, Colorado, and contained 780 mg Pb kg^{-1}.*

Sludge in diet %	Diet Pb	Pb in tissues		
		Kidney	Liver mg Pb kg^{-1} dry wt	Bone
0	0.5	0.9 a	0.2 a	0.8 a
4	29.0	12.2 b	3.3 b	3.7 b
12	80.0	15.8 c	4.6 c	11.0 c

Within a column, means followed by the same letter are not significantly different at the 5% level.

Table 14. *Effect of percentage of sewage sludge compost in diet on Pb residues in tissues of cattle which consumed the test diet for 180 days (Decker et al., 1980). The compost contained 215 mg Pb kg^{-1}, and high levels of Fe and Ca.*

Dietary composn %	Pb concentration					
	Diet	Faeces	Duodenum mg Pb kg^{-1} dry weight	Liver	Kidney	Femur
0.	6.0 a	14.7 a	2.81 a	2.36 b	3.96 ab	3.70 a
3.3	11.2 b	23.8 b	3.18 a	2.48 b	5.26 a	4.74 a
10.0	19.9 c	46.7 c	4.21 a	8.44 a	2.92 b	3.37 a

Within a column, means followed by the same letter are not significantly different at the 5% level.

compost containing lower amounts of lead (380 µg g^{-1}) had no significant change in their bone or liver lead concentration (Decker *et al.*, 1980). Cattle fed sludge compost containing 215 mg Pb kg^{-1} had little change in their femur lead (Table 14). This confirms the hypothesis that the nature of the soil matrix can greatly reduce the bioavailability of lead in ingested soil-like materials. This work also indicates that soil may adsorb lead so strongly that PbB is not increased until some threshold soil-lead concentration is exceeded. The threshold was found to be about 300 mg Pb kg^{-1} for sewage sludge compost ingested by cattle (Chaney *et al.*, 1989).

Many scientists have considered the bioavailability of lead in ingested soil and dust (Chaney *et al.*, 1984; Day *et al.*, 1979; Duggan and Inskip, 1985;

Ferguson *et al.* 1986; Gibson and Farmer, 1984; Harrison *et al.*, 1981; Thornton, 1986). Some have conducted chemical extractions to simulate conditions of the stomach, and found soil lead was very soluble (Chaney *et al.*, 1984; Day *et al.*, 1979; Duggan and Inskip, 1985; Ferguson *et al*, 1986; Harrison *et al.*, 1981). Although some argue that solubility means availability, the research work discussed above shows that soil components may adsorb stomach-solubilised lead at the pH of the intestine and thereby reduce lead absorption. This effect would cause increased lead absorption at higher soil/dust lead concentrations due to the saturation of intestinal binding sites. The reports discussed showed that a decreasing lead response slope results when increasing amounts of soil are ingested.

F. Potential Importance of Stomach pH on Absorption of Lead from Ingested Soil and Dust

Questions have been raised about which soil properties most reduce lead bioavailability and which experimental animal species might be the most appropriate models for determining the risk to infants/children from lead in urban soils, mine wastes, lead-ore concentrates, smelter wastes and lead-paint-contaminated soils. Because of the potential effect of stomach acidity on the rate of dissolution for PbS *etc.*, stomach pH was reconsidered. In particular, an evaluation was made of stomach pH and the effect of soil ingestion on soil pH of humans, pigs, and rats, because of the apparent importance of stomach pH in the dissolution of PbS and soil lead. Many Superfund sites involve mine wastes which appear to contain predominantly PbS.If PbS has low bioavailability to humans under normal environmental exposure conditions for the worst case in children, the cost of remediating these sites may be much reduced if the lead in the mine waste/soil is known to have lower bioavailability than that found in soils contaminated by smelter emissions, automotive emissions, or paint residues. These latter sources have been found to cause increased PbB in children exposed to soils when soil lead exceeds 500–1000 µg g^{-1} (USCDC, 1985;USEPA, 1986 Air Quality Criteria Documents; Duggan and Inskip, 1985). PbB in children exposed to PbS in mine wastes or ore concentrates appeared to have substantially lower response to this source than seen in other populations exposed to more soluble lead species in soil or dust (Middaugh *et al.*, 1989; Steele *et al.*, 1990).

Research has shown that PbS dissolution is dependent upon pH. The chemical solubility of PbS responds to both pH and particle size (Healy *et al.*, 1982; Roy, 1977). Because of the short incubation of food in the stomach, and possibly because of the pH buffering of food, mine wastes, PbS and larger particle size materials should not be expected to be dissolved in the stomach. PbS was found to be very much less soluble in human gastric juice than were lead carbonates and sulphates (Carlson and Woefel, 1913; Woefel and Carlson, 1914). Chamberlain *et al* (1978) fed PbS with food to human volunteers and found about 6-12% absorption.

Thus, the actual pH of the stomach contents during digestion of soil might be very important in assessing risk from soil ingestion. Researchers on microelement nutrition have considered the pH of the stomach and the duodenum in order to develop *in vitro* iron bioavailability assays (Miller *et al.*, 1981; Schricker *et al*, 1981; Reddy *et al*, 1988). It is difficult, but one must consider stomach pH under both the fasting condition and the effect of food (or soil) on the pH of the stomach contents. The number generally described as the pH of the stomach is the pH of gastric fluid secreted by fasting individuals. Actually, much is known about this because the importance of gastric fluid pH on ulcer development in humans. Hereditary, hormonal and dietary influences on gastric acid secretion cause pH to vary from 1.0 to 2.5 (or even as high as 7 when there is a poor ability to secrete stomach acid, e.g. in achlorhydria (Bezwoda *et al.*, 1978). However, as soon as food is ingested, the buffering capacity of the food causes the pH of the stomach contents to rise (Longstreth *et al.*, 1975; Malagelada *et al.*, 1976; Malagelada *et al.*, 1977; Malagelada *et al.*, 1979). Many of the techniques developed, and studies conducted, are concerned with excessive gastric acid secretion or sensitivity of stomach or duodenal tissues to stomach acid resulting from human ulcer disease. Compounds used to counteract ulcers inhibit acid secretion, and raise the pH of the stomach (*e.g.* Lucey *et al.*, 1989). Antacids also react with gastric acids to raise stomach pH. Soils could also cause the pH of the stomach to rise. The presence of $CaCO_3$, especially finely divided $CaCO_3$ in calcareous soils, but also neutral soils with higher cation exchange capacity, would cause a similar increase in pH of the stomach contents.

It is generally agreed that normal gastric fluid pH is 1–2 in rats, pigs and children. However, infant (pre-weaning, <24 days old rats have a high stomach pH (6–7), and the transition to a strongly acidic stomach pH is delayed compared with children (Takeuchi *et al.*, 1981). Johnson (1990), who had performed extensive research on gastrointestinal physiology with rats, noted that another likely source of misunderstanding about stomach pH results from the way we manage rats. The fasting rat stomach fluid pH is 1–1.5; however, the rat usually nibbles intermittently but continuously throughout the day, and much data about the rat stomach pH show a higher pH level because food is always present in the stomach. The human eats at set meal times, and in between meals accumulates a "basal" gastric fluid of pH 1–2 in the antrum of the stomach. When food is ingested, the pH rises to 5–6. The rat continuously secretes stomach acid, and secretion responds to several hormone activities. The human has a low basal secretion, but hormones significantly increase acid secretion when food is ingested or the stomach is distended. On a diet of rat chow, the rat stomach empties slower than human stomachs, but this may be an artifact of the digestibility of human foods compared with rat chow. Johnson (1990) notes that rat and human stomach pH levels are not that dissimilar, and that rats are a valid model for processes which are pH dependent such as PbS dissolution. Both secrete a solution which has a pH of 1 to 1.5.

Another source of information about stomach pH comes from the work of

Bates (1990) who has studied the bioavailability of food iron and tried to develop *in vitro* methods to assess the bioavailability of Fe. Cannulated miniature pigs were fed test meals, then the gastric fluid and duodenal fluid were sampled. A pinto bean meal caused stomach pH to be 5.1, 4.0, and 3.1 at 30, 60 and 90 minutes after the introduction of the homogenised slurry of pinto beans test meal (Reddy *et al.*, 1988). These pH levels are very similar to the results for adult humans reported by Malagelada *et al.* (1976; 1979).

These considerations indicate that the stomach pH of rat, pig, and human children are not different enough to justify the use of the pig rather than the rat in assessing bioavailability of lead in soils and mine wastes; but, because of the potential extreme public expense in remediating lead polluted urban soils and mine wastes, important principles of soil-lead bioavailability shown in rats may need to be confirmed in pigs and primates to establish the need for these public costs.

It is important to consider that food and soil can buffer the pH of the stomach to high levels (pH 6) greatly reducing dissolution of environmental PbS and/or soil-lead. Limestone in soil or mine wastes, or higher cation exchange capacity neutral pH soils might consume gastric acidity and thus allow the digesta to enter the small intestine without receiving the strong acid attack normally assumed to take place in the stomach.

VII. LEAD IN HUMAN HEALTH

This chapter relates especially to Steps B, E, H, K and N of the Phased Action Plan.

In this chapter the factors concerned with lead and health are discussed in terms of the population groups at risk and in terms of adverse health effects of lead. The resulting health criteria were then used to derive a target soil or dust lead guideline concentration.

A.*Population Groups at Risk for Adverse Health Effects of Lead*

Exposure to lead and its adverse health effects have been intensively studied, particularly during the past 15 years. These studies, which have been reviewed and critically evaluated in the report "Air Quality Criteria for Lead" (US EPA, 1986), have identified the foetus and young child population groups as those at greatest risk for adverse health effects of lead. In addition, analyses of the NHANES-II data (Mahaffey *et al.*, 1982) suggest that: (1) a modest increase in blood lead concentration (PbB) in middle-aged white males may be associated with a very small but statistically significant increase in blood pressure and, (2) there may be a 25% increase in PbB in post-menopausal women. A number of prospective studies are in progress addressing the issues involved, and they are discussed below.

1. Foetus

Paul (1860) was the first to report an increased incidence of spontaneous abortion and stillbirth in pregnant women with clinical manifestations of severe lead poisoning. This observation was confirmed by others during the next 50 years and led to the recommendation that women be excluded from the lead trades, a recommendation that has been accepted in many countries. Tissue analysis of the products of conception (Barltrop, 1969) reveal that lead freely crosses the human placenta and accumulates in the tissues of the foetus at a high rate during the third trimester of pregnancy. The concentrations of Pb in the various tissues of full term stillbirths were found to be equivalent to those reported in non-pregnant adult females (Barry, 1975). A number of studies have shown that maternal PbB and infant cord PbB are essentially equivalent at birth.

2. Child: Birth to 6 or 7 years.

Byers and Lord (1943) were the first to report that clinically acute lead poisoning during early childhood had lasting neurotoxic sequelae. Nineteen of the twenty children whom they followed through the early school years were excluded from school. They attributed this primarily to anti-social behavioural disorders, short attention span and sensimotor deficits. They remarked upon the fact that these children failed in school despite the fact that they had apparently normal intelligence quotients (IQ) as judged by global intelligence scores.

By the early 1970s (Lin-Fu, 1973), emphasis began to change from treatment to prevention and to the study of the effects of low level lead exposure in asymptomatic children. The work of Needleman *et al.* (1979) demonstrated, in a general population cross-sectional study of first and second grade school children, a significant 4 point reduction in verbal IQ, shortened attention span and a dose related increase in the frequency of unfavourable classroom behaviours. A recent follow-up evaluation of this cohort (Needleman *et al.*, 1990) suggests that these early developmental effects are reflected in later high school performance and academic success. A number of cross-sectional studies have since been reported and reviewed in detail elsewhere (U.S. EPA, 1986; Smith, Grant and Sors, 1989; Bellinger *et al.*, 1991). Some have confirmed the findings of Needleman and his group, while others have failed to find statistically significant differences. A number of the studies have been criticised on the basis that the number of subjects in each group were too small to achieve adequate statistical power or that the studies failed to allow for important confounding factors.

3. Prospective Studies

The limitations of the cross-sectional studies have led to the setting up of a number of prospective studies on lead absorption and its health effects. These are now in progress in the United States, the United Kingdom, Australia, Yugoslavia and Mexico. In these studies, women are enrolled during pregnancy and their offspring are followed longitudinally at least until school age. These studies are notable for the numerous covariates and potentially confounding variables that have been taken into account in the analyses of the data. The studies in Boston and Cincinnati in the United States and in Port Pirie, South Australia, are the most advanced at present. The findings have been extensively reviewed elsewhere (U.S. EPA, 1986; Davis and Svendsgaard, 1987; Smith *et al.*, 1989), and show a significant reduction in gestational age that is inversely related to cord or maternal PbB levels. This has been a consistent finding in most studies. Some, but not all, studies have shown a reduction in birth weight at blood lead levels greater than 12–13 µg dl^{-1}.

The Boston study (Bellinger *et al.*, 1987, 1990, 1992) consists of 249 children who were divided into 3 prenatal lead exposure groups based upon cord blood levels. These were, low (PbB 0–3 µg dl^{-1}), mid (PbB 5–8 µg dl^{-1}), and high (10–18 µg dl^{-1}). The average postnatal PbB in this upper middle class cohort of infants was 5–7 µg dl^{-1}. The investigators reported that infants in the high prenatal exposure group showed impaired mental development at least until 2 years of age, as indicated by the Bayley Mental Developmental Index. After correction for covariate and confounding variables, mean IQ differed by 7 points between the low and high cord blood lead groups. An inverse relationship was found between PbB at 24 months of age, but not cord blood lead, and cognitive abilities at 57 months of age. At this age, the mean performance of children with high prenatal exposure was indistinguishable from that of children with lower prenatal exposure. In some children who had PbB less than 10 µg dl[-1] throughout postnatal life, there was an improvement at 57 months of age, but it could not be determined whether this improvement resulted from recovery or compensation. This group found (after adjustments for potential cofounders) that higher blood levels at age 24 months were significantly associated with lower global scores on both the Weschler Intelligence Scale for Children, Revised (WISC-R) and the Kaufman Test of Educational Achievement (K-TEA) at 10 years of age.

The Cincinnati study (Dietrich *et al.*, 1985, 1986, 1990, 1993a, 1993b) is examining a cohort of 292 disadvantaged children who are regularly assessed on blood lead levels, neurobehavioural status and general health. The authors suggest that the effects of foetal lead exposure on early neurobehavioural outcomes is partly due to the lead related reduction in foetal growth and maturation, as noted above. Blood lead levels rose rapidly in exposed infants after 6 months of age, but at two years, no statistically significant effects of pre- or postnatal lead exposure on neurobehavioural status could be detected.However, at approximately 6.5 years of age, averaged lifetime blood lead concentrations in excess of 20 µg Pb dl[-1] whole blood were associated with deficits in the performance component of the WISC-R on the order of about seven points when compared with children whose mean blood lead concentrations were less than or equal to 10 µg Pb dL[-1] whole blood.

Ernhart *et al.* (1981, 1988) have followed a cohort of 260 disadvantaged urban children in Cleveland, Ohio. The authors concluded that lead exposure in the range examined had little effect on child development. Variables relating to the caretaking environment contributed most to the variance of cognitive measures up to and including the third year of life. It should be noted that this study was originally designed to investigate the impact of foetal alcohol exposure on child development and that 50% of the women recruited reported a history of alcohol abuse. Such women were considered ineligible for most other cohorts of lead-exposed children in which the various investigators took great care to assess alcohol consumption and its contribution to development. However, Ernhart controlled for this significant cofounder with the MAST score and other comprehensive diagnostic tests as discussed in Ernhart and Greene (1990) providing her study with a powerful screen for alcoholism related problems which other similar studies might lack.

Over 700 children residing in and around the lead smelter town of Port Pirie, Australia are also being studied (McMichael *et al.*, 1985., 1988). At two years of age, there was a weak inverse association between postnatal PbB and intelligence, however no relationship was found at age three. Intelligence measured at four years of age was inversely related to a linear integration of average PbB concentrations up to 3 years of age. As the integrated PbB concentration increases from 10 to 31 µg dl[-1], the general cognitive index decreased by 15 points, 7.2 points of which were attributable to lead after correction for covariates and confounders. This group has now been followed to seven years of age. Average blood lead concentrations, as determined from measurements at 15, 24, 36 and 48 months of age, show, at 7 years of age, a 4 – 5% reduction in intelligence quotient.

Cooney *et al.* (1989) followed a cohort of middle class children in Sydney, Australia, with 202 remaining at 4 years of age. No statistically significant correlations were found between mental or motor development at 3 or 4 years of age and prenatal or postnatal blood lead levels.

While most of the prospective studies are finding lead related developmental effects, the nature of the exposure patterns with which they are associated are not entirely consistent across all studies. Nevertheless, effects persisting to 0.5 to 10 years of age have now been reported in the Boston, Cincinnati and Port Pirie prospective studies which suggest that all socioeconomic classes of children are affected (Mahaffey, 1992). The findings in these studies are consistent with those reported in non-human primates (Needleman, 1992). Even so, much remains to be learned .regarding the relative importance of prenatal and postnatal exposure, the persistence of the effects, and their long term impacts on social and academic competency.

4. Organ Sensitivity

The primary target organs for lead are the central nervous system, the haematopoietic system and the kidney (US EPA, 1986). Effects on other organ systems have been reported, but occur only at very high levels of exposure. Lead has been shown to inhibit haem synthesis in every organ studied. Furthermore, each cell synthesises its own haemoproteins. The principal enzymes affected are porphobilinogen synthase (PBGS), usually known as delta-aminolaevulinic acid dehydrase (ALAD) and ferrochelatase. These partial inhibitions are associated with a pathognomonic constellation of biochemical changes, including *in vitro* inhibition of PBGS activity in peripheral blood, increased zinc protoporphyrin in erythrocytes and increased outputs of delta-aminolaevulinic acid (ALA) and coproporphyrin in urine in association with normal or slightly elevated outputs of porphobilinogen and uroporphyrin in urine. While the PbB threshold for inhibition of PBGS activity *in vitro* lies at a PbB of 5-10 µg Pb dl^{-1} or less (Chisolm 1985), the blood lead threshold for increasing zinc protoporphyrin is at 15-18 µg Pb dl^{-1} (Piomelli *et al.*, 1982; Hammond *et al.*, 1985). The effects of lead on the biosynthesis of haem are reversible, including lead-induced anaemia. Inhibition of haem synthesis in the developing erythrocyte in the bone marrow was identified as the critical or most sensitive adverse affect of lead (Nordberg, 1976).

As erythrocyte protoporphyrin levels are also increased in iron deficiency states, this established the usefulness of micro erythrocyte protoporphyrin tests for screening purposes. With few exceptions, 95% or more of the porphyrin in circulating erythrocytes is zinc protoporphyrin (ZnPP), which can be measured directly. In certain extraction procedures zinc is removed and the term "free" erythrocyte protoporphyrin (FEP) is used. The results are variously reported as µg per dl of whole blood or erythrocytes or per gram of haemoglobin. The nature of the dose response curve between blood lead is such however, that the test does not have the sensitivity and specificity for lead in blood at current levels of concern, namely PbB less than 20 µg dl^{-1}. It should also be noted that some authors (*e.g.*, Moore and Goldberg, 1985) believe that changes in the intermediates of the haem biosynthesis noted above do not themselves represent adverse health effects, and that therefore an adverse health effect would not occur until haemoglobin production itself has fallen.

In the light of more recent data on the neurotoxic effects of lead, it would appear that the developing nervous system is at least as sensitive, if not more sensitive to lead, than in haem synthesis in the bone marrow, an effect which is reversible. Furthermore, experimental data and studies in man strongly suggest that the neurotoxic effects of lead are not reversible. Various other metabolic and neurotoxic effects of lead as studied in experimental animals and man are beyond the scope of this text but are reviewed extensively elsewhere (US EPA, 1986; Needleman, 1992).

5. *Endogenous Factors Affecting the Susceptibility of the Foetus and Young Child to Lead*

Factors which render the foetus and young child more sensitive to lead than older children and adults relate primarily to the very rapid growth rate during this early period. Indeed, the nervous system has a growth rate more rapid than other tissues during the latter part of foetal development and early postnatal life up to about 6 years of age. The brain in infant experimental animals tends to accumulate and retain lead long after dosing with lead is stopped, a phenomenon not observed in mature animals. It has also been shown (Ziegler *et al.*, 1978) that human infants from birth to 2 years of age absorb approximately 50% of dietary lead, one half of which is retained. By contrast, adults absorb 8–12% of dietary lead, only a very small fraction of which is retained. Indeed, the data of Ziegler *et al.* (1978) indicate that an infant is in positive lead balance when the dietary intake of lead exceeds 5 µg Pb kg^{-1} body weight/day. Studies in human adult volunteers indicate that the absorption of lead is increased by a factor of 3 to 5 when administered in the fasting state. It should further be noted that infants have a diet composed primarily of milk. In experimental animals milk has been shown to increase the absorption of lead. Cow's milk, unless fortified, is also deficient in iron and copper. The bioavailability of zinc may also be reduced in cow's milk. The experimental data in animals indicate that deficiencies of these elements enhance the absorption and retention of lead (Mahaffey, 1981). The demands of growth render infants and toddlers highly susceptible to nutritional deficiencies. In summary, very rapid growth rate, particularly of the neural system, and the high rate of intestinal absorption and retention of lead are the principal factors which make the foetus and young child the population group at highest risk for over exposure to lead and its adverse health effects.

B.   *Populations at Risk for Exposure to Lead in Soil*

Infants and children, from birth to 6 or 7 years of age, constitute the group at greatest risk for exposure to lead in soil. Within this overall age range, children may be divided into two age groups; namely, 6–36 months of age and 37–72 months of age, based primarily upon developmental and behavioural considerations. Even so, it should not be forgotten that studies among older children living in proximity to stationary point sources of lead emissions, such as smelters, have also shown increases in PbB although the degree of increase has not been as great as it is in younger children similarly exposed (Landrigan *et al.*, 1975; Yankel *et al.*, 1977; Roels *et al.*, 1980). The hand to mouth route of lead in soils and interior household dust has been well documented as a major pathway of environmental lead into the bodies of young children (Sayre *et al.*, 1974; Roels *et al.*, 1980; Bornschein *et al.*, 1986).

1. *Children 6-36 Months of Age*

Until approximately 6 months of age, infants tend to spend virtually all of their time in cribs (cots). Between 6 and 12 months of age, infants begin to scoot, crawl and walk, thereby enabling them to move freely about the home during which time they become more highly exposed to lead in interior household dust. A portion of this dust represents lead in exterior soil tracked into the home, as well as that which may be blown in through open windows and doors. During this age period, infants and toddlers tend to spend 80–90% of their time indoors.

The most important factor is the prevalence of hand to mouth activity as a normal developmental component of behaviour in this age range. Virtually all children suck their thumbs and fingers during their first year of life. Sucking fingers occurs as a result of the sucking reflex. Barltrop (1966) noted its occurrence at 12 months of age in 80% of children studied, as determined on the basis of 24 hours and 14 days recall by parents. After 12 months of age, thumb sucking and finger sucking tend slowly to diminish over the next five years, at the end of which, perhaps, only 20–30% of children are reported to show this activity. Between 12 and 72 months of age, thumb sucking generally occurs in relation to fatigue, boredom, illness, punishment and other frustrating situations to which the child responds by regressing to a more infantile type of behaviour. After 5–6 years of age, finger sucking should be considered as evidence of emotional immaturity (Harper and Richmond, 1977). Among older children, particularly males, playing in the dirt and a disregard for cleanliness go hand in hand. Thus, exposure to lead in soil can persist well into the school years. Indeed, the study of Roels *et al.* (1980) was based primarily on older school-aged children and environmental data obtained by measuring lead in soil in school yards.

It is normal for a child to mouth foreign objects during infancy. If they ingest non-food items such as any item from the floor, dirt, plaster, wood, sand etc., the habit may be defined as *pica*. During infancy, this represents oral exploration of the environment. Why children exhibit this habit is unknown, although a dietary deficiency of iron has been proposed, but not substantiated. Lourie *et al.*(1963)

considered the "absence of mothering" as an important factor in the aetiology of pica. "Absence of mothering" might be due to the fact that the mother was out at work, preoccupied with younger infants, mentally disturbed or an abuser of alcohol or drugs. In the study of Barltrop (1966) pica or ingestion of non-food substances was observed to occur at about one-half the frequency of mouthing during the age period from 12–72 months. Some mentally deficient children may persist in the habit of pica throughout childhood and well into the adult years. In more recent studies on lead exposure in preschool age children the Caldwell HOME inventory has been used to assess the role of care giving in the home on mental development. Several investigators have found that three sub scales of the HOME in particular (Maternal Involvement with Child, Provision of Appropriate Play Material, and Emotional and Verbal Responsivity of Mother) were negatively correlated with cumulative lead as indexed by serial PbB measurements. It was noted that, even within the same socio-economic class, a wide variation in the quality of care giving was found. These findings suggest that infants and toddlers receiving inadequate social and physical stimulation may indulge in greater amounts of hand to mouth activity than those similarly exposed to lead in dust and soil but for whom the quality of caring was higher (Schroeder, 1989). General cleanliness of the home also has been shown to influence PbB levels in children (Yankel *et al.*, 1977).

2. *Children 37–72 Months of Age*

Between 3 and 6 years of age, the habits noted above persist but decrease in prevalence and frequency. Also, growth rate has decreased substantially. Conversely, children in this age range will tend to spend more time outdoors where they can be exposed directly to lead in soil in their play areas, particularly if their play areas are bare soil.

C. *Definitions of Acceptable Blood Lead Concentrations*

Measurement of the concentration of lead in whole blood provides an indicator of the internal dose of lead and has served in epidemiological surveys as the most widely used indicator of lead absorption for the past 20 to 30 years. The total amount of lead in whole blood at any point in time is the sum of both recently absorbed lead and lead absorbed in the past. For example, an isolated brief episode of sharply increased lead absorption will sharply elevate blood lead concentration for a short period of time. Lead is stored primarily in bone from which it is slowly recycled back to the blood over a long period of time and remains the most useful index of lead exposure and absorption for the purposes of epidemiological surveys.

Lead can and has been measured in urine and hair. There may be wide variation in concentration of lead in urine so that urine lead measurements are not useful for epidemiological surveys. Lead may be adsorbed on to hair so that a hair lead measurement does not necessarily represent what has been absorbed into the body and then incorporated into the hair. This measurement is considered of no use for epidemiological purposes. The use of the calcium disodium EDTA mobilisation test for lead has been largely limited to clinical research. Furthermore, this test may be hazardous since experimental studies indicate that a single dose may elevate the concentration of lead in brain and liver (Cory-Slechta *et al.*, 1987). Newer techniques for the measurement of lead in bone as well as the measurement of lead in shed deciduous teeth still fall within the realm of research.

All the clinical data on the effects of lead on early neurodevelopment have used either blood lead concentration or the concentration of lead in shed deciduous teeth as the index of the lead dose. Although the need for a comprehensive soil lead survey may initially be proposed from an isolated measurement of lead in hair, an elevated FEP test or some other test, no decision should be reached until blood lead data are available.

1. *Historical Lowering of Acceptable Blood Lead Concentration*

Historically, an acceptable blood lead concentration has been defined as that concentration below which adverse health effects, as understood at that time, were not likely to occur. The development of the colorimetric dithizone technique for measuring lead in biological tissues and fluids, including blood, made measurements feasible for the first time on a reasonably wide scale. The "dithizone era" lasted from the early 1930s until about 1970 when it was replaced by atomic absorption spectrophotometry (AAS) and anodic stripping voltammetry (ASV). Micro AAS methods are the most widely used ones today.

The dithizone method was cumbersome, difficult, and most laboratories required 10–20 ml of blood for a single analysis. By contrast, modern ASV techniques require only 100 microlitres of blood and micro AAS methods require far less. Furthermore, it is now possible to use samples in which lead has been determined by isotope dilution-mass spectroscopy (ID-MS), the ultimate reference method for lead and for the primary standardisation of alternate techniques. Also, quality assurance and quality control methods have become highly developed during the past 15 to 20 years which gives greater assurance of accurate results.

Given the large amount of blood required for analysis during the dithizone era, it is not surprising that blood specimens were not usually taken unless there was a strong clinical suspicion that the patient had lead poisoning. Furthermore, medical interest was concerned primarily with the acute clinical disease.

Papers published in the literature up until about 1970 dealt primarily with the diagnosis, treatment, and sequelae of acute clinical lead poisoning. Thus, the upper limit of acceptable blood lead concentration in adults until about 1970 was 80 µg Pb dl^{-1} whole blood. This limit was chosen because acute lead colic or other adverse effects almost never encountered at lower concentrations. In children, up until 1960-1965 the upper limit of acceptable blood lead concentration was 50–60 µg Pb dl^{-1} whole blood. This was based largely on the observation that one did not encounter, on X-rays, bands of increased density at the growing ends of the long bones (the so called "lead line") at lower blood lead concentrations. In contrast, early, nonspecific clinical manifestation such as irritability and anorexia might be encountered above this level. Indeed during this era, due to the lack of availability of blood lead measurements, the diagnosis of lead poisoning in children was often based on X-ray findings and the presence of basophilic stippling of erythrocytes, both rather insensitive indices of lead absorption.

It has long been known that lead disturbs haem synthesis as manifested by increased concentration of protoporphyrin in circulating red blood cells and increased output of coproporphyrin and delta-aminolaendinic acid in urine. Reliable quantitative techniques for coproporphyrin and delta-aminoleminic acid in urine were developed in the 1950s and used rather widely in the 1960s in studies to determine the dose-response and dose-effect relationships for these metabolic evidences of toxicity due to lead. It became apparent that the blood lead threshold for these responses in both children and adults was at a PbB level of approximately 40 μg Pb dl^{-1} of whole blood (NAS/NRC, 1972). At the same time, studies in children without exposure to lead beyond that found in usual food, water and air at the time did not exhibit PbB in excess of 40 μg Pb dl^{-1} whole blood. In 1970, the Surgeon General of the United States proposed 40 μg Pb dl^{-1} whole blood as the upper limit of normal or acceptable blood lead concentration (US Dept. Health, Education and Welfare, 1971).

By the late 1960s it became apparent that chelation therapy, although effective in reducing mortality from acute lead encephalopathy, did not result in any dramatic reduction in the occurrence of permanent CNS sequelae in children with recurrent episodes of clinical lead poisoning. Interest, therefore, shifted, together with an awakening of social consciousness in the mid 1960s, from case finding and treatment to prevention of lead toxicity. The critical effect concept, as fully described in the report of the Subcommittee on the Toxicology of Metals of the Permanent Commission and International Association of Occupational Health, provided the scientific rationale and practical approach for the prevention of lead toxicity (Nordberg, 1976). Under this concept, if the "critical" or earliest measurable adverse health effect can be identified and effective action is undertaken on this basis, then later and more serious effects can be prevented. Disturbance of haem synthesis in the bone marrow was identified in this report as the critical effect of lead. At the time, the most sensitive measure reported was *in vitro* inhibition of the activity of delta-aminolaevulinic acid dehydrase in circulating red blood cells. However, the significance for health of this *in vitro* measure was uncertain.

At about the same time, micro methods for the measurement of zinc protoporphyrin in circulating erythrocytes either as zinc protoporphyrin or as "free" erythrocyte protoporphyrin became available for screening purposes. In 1978, the US Centers for Disease Control recommended that the upper limit of acceptable blood lead concentration be lowered from 40 to 30 μg Pb dl^{-1} whole blood.

2. Current Reference Values.

It is clear that blood lead concentrations in the general population have been declining for at least 20 to 30 years both in the U.S. and the U.K. The data prior to 1970 are based upon the use of the dithizone method, so that some of the decline may be attributable to the concomitant lack of sensitivity and precision. However, this does not appear to be true during the atomic absorption era as there have been no significant changes in methodology over the past 15 years or so. In the U.K., it is estimated that blood lead concentrations have been decreasing by 4–5% per year during the past decade (Quinn and Delves, 1989). During 1986 the geometric mean PbB in children and women in the U.K. was approximately 8 µg Pb dl^{-1} whole blood. Decline in PbB in the United States has been greatest during the past decade. The NHANES-II data indicate that between 1978 and 1980 mean PbB in the United States decreased from 15.9 to 9.6 µg Pb dl^{-1} of whole blood (Annest *et al.*, 1983). It is currently estimated that mean PbB may be approximately 6 µg Pb dl^{-1} whole blood in the general population in the United States (ATSDR, 1988). In the United States this decline has been attributed to sharp reductions in both air and food lead.

At present the World Health Organization (WHO) recommendation for acceptable blood lead concentration is no more than 2% of the population with a PbB greater than 20 µg Pb dl^{-1} of whole blood. The latest recommendation (1991) of the U.S. Centers for Disease Control in the United States is that the upper limit of acceptable PbB should be 10 µg Pb dl^{-1} whole blood. This limit is based upon publications during the past seven years since the last CDC statement, which indicates that the risk in groups of children for long-lasting adverse neuro-behavioural effects increases as average blood lead concentration during the preschool years rises above the 10 µg Pb dl^{-1} whole blood. In this new statement, the erythrocyte protoporhyrin test is dropped as a primary screening tool because of its insensitivity and the current opinion that the critical organ in the foetus and young child is the developing nervous system and not the haematopoietic system, as previous data indicated. In the 1991 statement, the CDC attempt to accomplish two goals which are not entirely compatible: (1) to make a general statement to provide guidance for governmental agencies to reduce exposures such that blood lead levels will not exceed 10 µg Pb dl^{-1} whole blood in young children and (2) to provide guidance to health care providers in the management of individual children. The various levels above 10 µg dl^{-1} whole blood in this statement are keyed to the urgency with which individual children should be followed up. A third epidemiological purpose is to collect data to identify communities with the higher population mean blood lead levels. In the management of individual cases of increased lead absorption, it should be emphasised to parents that lead is one of several variables which will determine an individual child's ultimate cognitive abilities. Perhaps the most important of these is the care that the child receives after birth. The CDC's 1991 statement does not recommend specialised case management until blood lead concentrations exceed 20 μg Pb dl^{-1} whole blood.

In summary, the limits of acceptable PbB have changed substantially during the past 50 years. During each decade such limits were generally set on the basis of what was perceived at the time to be a significant adverse health effect. Initially, the aim was the prevention of acute clinical disease. More recently concern has centred on the neurobehavioural effects of lead exposure during foetal life and early childhood. The limits have also changed in relation to improving technology which has permitted the measurement of lead and its various adverse health effects at lower and lower levels as the sophistication of technology advanced. Thus, during this fifty year period one has seen a lowering of the upper limit of acceptable blood lead concentration in children from 60 μg Pb to 10 μg Pb dl^{-1} whole blood.

D. *Other Sources of Lead*

For purposes of classification, environmental sources of lead may be divided into three groups according to the concentrations of lead likely to be found in each of the sources: (a) low (or baseline) dose; (b) intermediate dose; and (c) high dose. In general, intermediate dose sources are associated with moderate increases in PbB in children up to 50–60 μg Pb dl^{-1}. Such children are asymptomatic. High dose sources, while often associated with similar increases in PbB, are also associated with much higher PbB levels (greater than 80–100 μg Pb dl^{-1}). In the latter group, acute clinical symptoms are likely to be found. Fatalities have also been reported in relation to some high dose sources. For a particular source the range of PbB levels found in groups of children that have been studied may span across all three classification groups. For example, the amount of lead- bearing particulates carried from workplace to home on clothing may vary widely so that some children are symptomatic while others show little effect (Baker *et al.*, 1977).

1. *Low (baseline) Dose Sources*

The general population is exposed to small amounts of lead in air, food and water. Lead in air and food has decreased dramatically in the 1980s in the United States (US EPA, 1986; ATSDR, 1988). The concentration of lead in drinking water varies widely around the world. In areas of the world where the water is low in total dissolved solids (<100 mg l^{-1}) and with an acidic pH soluble lead appears to depend primarily upon the presence of lead in pipes and solder in the distribution system. Such waters are generally described as "plumbosolvent" and "aggressive". This phenomenon has been intensively studied in northern England and Scotland (Moore *et al.*, 1985). When discovered, such water supplies can be treated at source by additions of lime to reduce the plumbosolvency of the water. At the delivery point (within the home) old lead pipes should be removed and replaced with plastic ones.

2. *Intermediate Dose Sources*

Sources in this group are generally those which contribute to the lead content of interior household dust and in some circumstances children's play areas. Included in this group are lead-bearing particulates brought into the home on the dirty work clothing of lead workers (Baker *et al.*, 1977; Chisolm, 1978). Removal of lead paint, particularly if flame gas torches, heat guns and mechanical sanders are used, can greatly and acutely increase interior and exterior lead bearing particulates. These practices have been associated with clinical illness in both workers and exposed children (Feldman, 1978; Farfel and Chisolm, 1990; Marino *et al.*, 1990). Similar increases occur when amateurs and residents ("do it yourself" enthusiasts) carry out these procedures unaware of the hazards involved (Fischbein *et al.*, 1981; Inskip and Atterbury, 1983). Even in the absence of paint removal work, interior household dust lead tends to be higher in older housing due to the weathering and chalking of old lead based paints. Cottage or home industries and hobbies involving the making of pottery, other ceramic ware and art glasswork can lead to rather high concentrations of particulate lead in the household dust particularly in the areas of the home in which these activities are carried out.

The fallout from primary or secondary smelter emissions can heavily contaminate the local area. In some areas water lead levels may be particularly high and may be associated with moderate increases in blood lead concentration. When these situations are investigated, the range in PbB concentrations found may be quite wide and include some individuals with blood lead concentrations high enough to be compatible with early clinical lead poisoning.

3. *Specific and Unusual High Dose Sources*

This group includes sources in which lead is more concentrated and with which cases of clinical lead poisoning, including fatalities, have been identified. Indeed, a number of the unusual sources have only come to light following the study of individuals with severe acute clinical plumbism. Within this group, the ingestion of lead based household paint is clearly the main source of serious lead poisoning in children, particularly those exhibiting pica. Such paints may contain 1–70% lead so that tiny bits may contain a highly toxic dose of lead. For example, paint flakes weighing 10 mg (about the size of a matchhead) contain approximately 10% lead (1,000 µg); if such flakes are eaten over the course of a few months they can lead to serious clinical disease. In the United States it is estimated that 52% of the current housing stock contains lead pigment paints on exposed residential surfaces (ATSDR, 1988). Paint in a defective condition constitutes an immediate and serious hazard.

Acidic beverages such as fruit juices, cola drinks, coffee and wine can leach substantial quantities of lead from crystal and improperly lead-glazed ceramic ware. Children have swallowed items made of lead such as curtain weights, fishing weights, shot, jewelry coated with lead to simulate pearl, and jewelry with a lead base. If these items remain in the stomach the lead will slowly dissolve.

The severity of the case will depend on how long the item remains in the stomach. The use of lead-contaminated health foods (usually calcium supplements) has also led to serious disease as has the use of herbal medicines from China, other parts of Asia and Mexico. Water stored in lead-lined cisterns on rooftops, and rain barrels used as a source of drinking water in close proximity to lead emitting plants, such as primary or secondary smelters, can produce severe disease as has the burning of battery casings in the home for heating and cooking. "Sniffing" (inhaling for pleasure) of leaded petrol has produced lead encephalopathy (Chisolm and Barltrop, 1979; Chisolm, 1985).

E. *Evaluation of Data From Survey by Follow up on Case Studies*

In any survey it is likely that more than one source of lead will be found in a given child's environment. It is rare to find a group in which soil is the only significant source of overexposure to lead. Those who conduct surveys have an ethical responsibility to see that the subjects receive appropriate medical and environmental follow-up. While this need not be done by the survey group it is still the survey group's responsibility to refer affected individuals to an appropriate authority. Analysis of the data from the survey may indicate that there is no significant over exposure to lead in soil. On the other hand, PbBs may be log normally distributed and significantly related to lead in soil which in turn would suggest that lead in soil is the major source of environmental lead for the group. If the distribution of PbB is greater than the range of PbB acceptable in a given community, a decision may be made to reduce overexposure to lead in soil. When this is done, there is a further obligation to do a follow-up survey including follow-up PbB measurements to evaluate the effectiveness of the steps undertaken under that decision.

The children who have participated in the study should also be evaluated as individuals. As a general rule, human research may be carried out only after approval of the appropriate human research or ethical committees. Ethical considerations require that research involving children may be conducted only if it is potentially beneficial to children. For this reason, data for each child must be evaluated on an individual basis not only for exposure to lead in soil but also to one or more additional sources of lead. Review of the questions asked should help to identify other sources such as lead bearing particulates borne into the home on the clothing of lead workers, cottage or home industries or home hobbies, lead contaminated water supply or defective lead-based paint. Ethnic groups known to use herbal medicines and/or lead-bearing cosmetics require evaluation from this point of view. Blood lead values well beyond the log normal distribution of blood leads in the group (statistical outliers) suggest either severe pica for paint chips or high lead dust with or without mental retardation.

The nutritional status of the group is likely to be established to some extent by the questionnaire. For example, FEP values significantly higher than those expected for a given PbB value strongly suggest iron deficiency usually due either to inadequate nutritional intake or chronic blood loss. In any event, all children who show either PbB or FEP levels beyond the acceptable limit should be referred for a complete medical investigation and then therapeutic intervention if that is deemed appropriate. The basic aims of intervention are usually to improve nutrition and reduce exposure. Steps needed to reduce exposure will need to be fitted to individual circumstances. Even if the physician to whom the child is referred elects to use chelation therapy, such therapy will only be of benefit in the long run if the sources of overexposure to lead in the child's environment are identified and effectively reduced.

F. Use of Health Criteria in Deriving a Target Soil/Dust Lead Guideline Concentration

1. Choice of model
A single guideline value for lead in soil to protect young children was considered unrealistic for a number of reasons. Various levels of blood lead concentrations are used as health standards around the world and these levels are changing as different criteria and effects of lead are considered. The environment of the population at risk can vary widely, from urban dusts derived from automotive emissions and leaded paints to smelter emissions, old mining areas, waste disposal sites or other sources. The population at risk can itself vary, such as in those situations where there is a high proportion of young children, or a retirement home for the elderly, or vacant land proposed for development: each represents very different risks. Because of these considerations, the soil/dust guideline was established as a dynamic concept or relationship based on a mathematical formula, in order to allow for a variety of environmental situations and regulatory criteria. A number of recent papers have discussed modelling techniques applicable to multiple source exposure to lead (Kneip et al., 1983; USEPA, 1986, 1988; Hoffnagle, 1988; Marcus and Cohen, 1988), and are discussed below. Alternate models can of course be used within the framework of this document according to the data available and any other priorities. In the last analysis, no matter what model is used, the blood lead-soil lead ratio will probably have to be determined in each situation on a site-specific basis. This is because so many variables, which may differ from one situation to the next, are involved. It should also be noted the slope of the relationship between blood lead and soil lead may vary considerably. The relationship may be non-linear over a soil lead range of $0 - 2,000$ μg g^{-1}. Bornschein et al. (1989) reported a slope of 6.2 for soil lead contents from 0 to 1,000 μg g^{-1}, while the slope decreased to 0.76 as soil lead increased from 1,000 to 2,000 μg g^{-1}

In the model used here, blood lead concentration is equated to a baseline level plus an increment resulting from exposure to lead in soil or dust. This may vary from one situation to another, based on differences in, e.g., food lead, water lead and social customs. The model takes account of the chosen blood lead guideline or target concentration and the degree of protection required in the population. The slope of the blood lead – soil lead relationship used in calculating increase in blood lead over a baseline value, and hence the soil guideline, can vary depending on a variety of factors, and this response can be adjusted for a given situation and modified as more data become available.

The relationship derived is

$$S = \left(\frac{\dfrac{T}{G^n} - B}{\delta} \right) 1{,}000$$

where **S** is the *soil or dust guideline*, a geometric mean concentration in g Pb per gram of soil or dust. **T** is the *blood lead guideline* or target concentration, in μ g Pb dl^{-1} whole blood. **G** is the *geometric standard deviation of the blood lead distribution*, typically it lies in the range of 1.3 to 1.5, but it may be higher for groups exposed to multiple and heterogeneous sources, *e.g.*, tailings piles in old mining areas, or soil contaminated by leaded paints. **B** is the *background or baseline blood lead concentration in the population from sources other than soil and dust*. Data from an appropriate control group would be ideal, *i.e.*, a group matched not only for population characteristics, but also for similar lead exposure from all sources except soil and dust. If there are appreciable contributions from other sources such as smelter emissions or leaded paint, these must be measured or estimated for addition to the baseline value. If these data are not readily available, any proposed investigation should evaluate the contributions from other suspected sources. **n** is the *number of standard deviations corresponding to the degree of protection required for the population at risk*, and would normally follow from the way in which the blood lead guideline T was defined. For example if 95% of the population should have blood lead concentrations less than 20 μg dl^{-1} then n can be obtained from standard statistical tables, and some representative values are given in Table 1 for different percentages of the population desired to be below the target blood lead concentration. δ *(delta)* is the *slope or response of the blood lead - soil(dust) lead relationship* and has the units of μg Pb dl^{-1} blood increase per 1,000 μg Pb g^{-1} increment of soil or dust lead.

2. Factors Affecting delta

The major uncertainty is the value to use for delta, the response of blood lead to increasing soil or dust lead concentration. A number of studies giving information on this relationship were considered, but it is not our intention to review them here, as most have already been considered in detail by the USEPA (1986) and by Duggan and Inskip (1985). These reviews also present a considerable number of other references relating the importance of dust and soil exposure to young children. A list of the papers considered is given in Table 15 along with details of the populations studied, the range of soil, dust and blood lead concentrations observed, and the estimated slope of the soil/dust - blood lead relationship. However, a number of observations need to be made, which bear on the choice of a value of delta to be used in this guideline model.

The range of slopes reported is wide, from 0.9 to 9.0 μg Pb dl^{-1} blood per

Table 15. *Studies relating blood lead and soil or dust lead concentrations.*

Study	Area[1]	Soil/Dust Conc.[2] μg Pb g⁻¹	Blood lead[2] μg Pb dl⁻¹	Age years	Number	Est. Slope g Pb dl⁻¹ blood per 1,000 g g⁻¹ soil or dust	Review[3]
Bornschein et al. (1989) Telluride, Co	M	172S 281 --567D	6	<6	94	2.2	
Moffat (1989) Dumfries, Scotland	M	213 – 69025 320 – 1570D	10 – 18	<12	37	1.2	
Phillips et al. (1989) Herculaneum, MO	M	70 --2258S 170 --2080D	7— 22	1 – 5	229	2.2	
Rabinowitz and Bellinger (1988) Boston, MA	U	702S (7–13, 240)	6	0.5 – 2	195	0.9	
Laxen et al. (1987) Edinburg, Scotland	U	500D (48–13, 600)	11 (−34)	8 – 9	495	1.9	
Milar and Mushak (1982) Raleigh, N.C.	B	250 – 3000D	18 --44	1 – 4	47	9.0	D
Reeves et al. (1982) Aukland, N.Z.	U	24 – 842S	12 – 19	1 – 3	195	5.0	D
Stark et al. (1982 New Haven, CT	U	230 – 1330S 160 – 630D	27	0 – 1 2 – 3 4 – 7	153 334 439	2.2 2.0 0.6	E
Roels et al. (1980) Belgium	S	112 - 2560D	9 – 25	10 --14	148	2.1, 3.5	D, E
Angle and McIntire (1979, 1982) Omaha, NE Trail, BC	U	81 – 339S 211 --479D	23 – 30	1 – 18	831	4.0, 6.8	D, E
Neri et al. (1978) Schmitt et al. (1979)	S, U	225 – 1800S	19 – 29 6	1 –3	87 103	7.6, 8.5 4.6, 7.2	D, E
Watson et al. (1978) Vermont	B	718 – 2239D	21 –32	1 –6	59	6.8	D
Baker et al. (1977) memphis, TN	S	500 – 5500D	22 –68	1 – 6	32	8.6	D
Yankel et al. (1977) Walter et al. (1980) Silver Valley, ID	S	400 –7500X	21 –66	1 – 9	860	1.1	D, E
Barltrop et al. (1975) Derbyshire, England	M	420 – 13970S 530 – 2580D	21 – 29	2—5	82	2.3	D, E
Galke et al. (1975)	U	173 – 1400S	32 – 43	<5	187	2.5	D, E
Shellshear et al. (1975) Christchurch, N.Z.	U	150 – 1950S	18 – 25	1 – 5	68	3.9	D, E
Roberts et al. (1974) Toronto, Canada	S, U	99 – 1715S 713 –1550D	17 – 27	0 – 14	1125	6.0	D

[1]. Area: B - battery plant, M - mining area, S - smelter, U - urban.
[2]. Soil (S), Dust (D) and Blood lead concentrations: means given low and high lkoad areas; otherwise, range of individual results.
[3]. Review: study reviewed by D - Duggan and Inskip (1985) or E - USEPA (1986).

1,000 µg Pb g^{-1} in soil or dust. Because of differences in design, the studies are not readily comparable. Some of the children studied were obviously exposed from multiple sources, and few of the studies measured all of the major sources of exposure either for individuals or adequately for the groups investigated. The age range of children was large, from less than 1 to 18 years of age, and widely differing environments were studied - smelter and battery plant exposure, old mining districts, and urban areas. Many of the baseline or background blood lead concentrations were considerably elevated by today's standards. In spite of these caveats, choices must be made, and the major factors to be considered in choosing a slope would be:

(a). The age distribution of children in the population at risk -two year old children exhibiting frequent hand- to- mouth activity would be expected to have a higher d than teenagers;

(b). The physical availability of dust and soil to the child. For example, is the garden or yard grass covered, or is it a dusty site? What is the degree of cleanliness of the home?

(c). The bioavailability of lead in dust and soil. This can vary with lead concentration in the dust or soil, the adsorption capacity of the soil/dust for lead, the chemical species of lead present (mine spoil compared with urban or smelter polluted soils and dusts), age of deposit, other soil components etc. The reader is referred to Sections V and VI for a more detailed treatment.

(d). The cultural/ethnic differences, such as parental supervision, time spent in/outside home, degree of clothing covering body surfaces, pica and mouthing habits etc.

3. *Choice and Use of delta*

In the review papers cited above, Duggan and Inskip (1985) chose to use a value of 5 for delta, being the average of the studies reviewed. The EPA review placed particular emphasis on the results of one study (Stark *et al.*, 1982) in that it provided good data for slope estimation, as well as providing data for both soil and house dusts for young children. These data indicate a delta of about 2 µg dl^{-1} per 1,000 µg Pb g^{-1} dust or soil.

Recent studies by Bornschein *et al.* (1989) and by Laxen *et al.* (1987) also indicate a delta of about 2, while Rabinowitz and Bellinger (1988) report a value of 0.9 for well maintained middle class neighbourhoods. Marcus and Cohen (1988) suggest a value of 2 as the most likely value, this being the median of the values reported in the papers noted. A more recent review of the literature by Steele *et al.* (1990), who made particular reference to mining areas, generally supports the above, and also gives a useful discussion of bioavailability.

It would appear that a value in the range of 2–5 would be appropriate for most situations. However, this should be adjusted in light of particular knowledge for a given situation. Low values of delta would tend to relate to groups with:(1) older children; (2) well maintained vegetative cover; (3) mine tailings (poor bioavailability);(4) cleaner homes and more frequent handwashing; or (5)

Table 16. *Variation of soil lead guideline with target blood lead concentration and degree of desired protection.*

Target PbB		Soil Lead Standard for % of Population <Target PbB			
μg dl-1	50%	95%	98%	99%	99.9%
10	3,000	880	500	300	-
15	5,500	2,300	1,860	1,400	700
20	8,000	3,750	3,000	2,600	1,600
25	10,000	5,200	4,250	3,700	2,500

Assumptions: $\delta = 2$, Background PbB = 4 g dl-1, GSD = 1.4

heavier textured soils. Conversely, higher values of delta would tend to be found in groups with: (1) children of peak lead absorbing and soil ingesting age, 18–24 months;(2) dusty conditions, sparse vegetative cover (*i.e.*, bare soil) ;(3) homes with poor cleanliness and infrequent handwashing; (4) soil lead sources with slight soluble lead salts such as automotive and stack emissions or well oxidized and more soluble sources including exterior paint; or(5) light textured or low organic matter soils. There is some inconsistent evidence that blood lead may increase more rapidly at low dose levels, which would mean than delta might not be constant over the entire concentration range encountered. Thus Rabinowitz and Bellinger (1988) and Bornschein *et al.* (1989) found higher values of delta at lead concentrations less than 1,000 μg Pb/g soil. The major influence on a value for delta may in fact be child activity, rather than any characteristics of dust or soil.

Tables 16 – 19 give examples of soil guidelines derived for different target blood lead concentrations by varying the other parameters in the model. Blank entries in the tables indicate where the model predicts negative guideline concentrations, where for example, lower blood lead target concentrations, T, or higher baseline blood lead concentrations, B, are selected. In these situations, a zero guideline would result, and consideration should be given to reducing other sources of lead exposure besides soil and dust lead and to re-assessing the public health criteria used and the   control measures available. Subsequent sections suggest a means of setting a guideline for undeveloped land, give examples of calculations for differing situations and health criteria, and also setting a guideline based on protection of the most sensitive individual rather than a population based guideline as used here.

It cannot be stressed too strongly that these are theoretical calculations, and if adequate blood lead data are or become available, they should, of course, take precedence in any decision making process. The availability of any relevant value for blood lead data can reduce the uncertainty arising from use of the model. Data from a suitable control group can reduce the uncertainty about B, the

Table 17. *Effect of variation in d and target pbb on soil lead guideline.*

Target PbB	δ, (g Pb dl^{-1} blood) (100 mg Pb kg^{-1} soil)$^{-1}$			
μg dl^{-1}	1	2	4	8
10	600	300	150	75
15	2,900	1,400	700	350
20	5,200	2,600	1,300	650
25	7,500	3,700	1,850	925

Assumptions : 99% of population PbB-T; GSD = 1.4,
Background PbB = 4 g dl^{-1}.

Table 18. *Effect of variation in the geometric standard deviation (gsd) of the pbb distribution on soil lead guideline.*

Target PbB	Geometric standard deviation					
μg dl^{-1}	1.3	1.4	1.5	1.6	1.7	1.8
10	720	300	-	-	-	-
15	2,100	1,400	930	520	190	-
20	3,400	2,600	1,900	1,350	920	560
25	4,800	3,700	2,900	2,200	1,650	1,200

Assumption : δ = 2, Background PbB = 4.0, 99% of population Pb-T.

Table 19. *Effect of variation in background pbb on the soil lead guideline.*

Target PbB	Background PbB g dl^{-1}				
μg dl^{-1}	2	4	6	8	10
10	1,300	300	-	-	-
15	2,400	1,400	450	-	-
20	3,600	2,600	1,600	600	-
25	4,700	3,700	2,700	1,700	700

In this case, 'background' could include other particular sources of lead exposure.
Assumption : d = 2, GSD=1.4, 99% of population

background or baseline blood lead from other sources, and data from the population at risk should provide more reliable estimates of delta, as well as of G, the geometric standard deviation. Any guideline chosen should be modified in light of future research, and, in particular, from any investigation initiated as a result of using any guideline established using the relationships discussed here. New evidence on health effects and revised public health criteria may also require a reassessment of any chosen guideline.

Other models of the blood lead-soil lead relationship may, of course, be used instead of the one presented. The important thing to remember is to make use of such a model within a properly developed phased action plan such as that described here.

4. Modelling the Blood Lead/Soil Lead Relationship

A number of recent papers have discussed modelling techniques applicable to multiple source exposure to lead, (Kneip *et al.*, 1983; USEPA, 1986, 1988; Hoffnagle, 1988; Marcus and Cohen, 1988), and should be consulted for additional discussion.

The 'disaggregate' modelling approach uses empirical relationships between blood lead concentrations and the concentrations of lead in the various media contributing to lead exposure. The slopes or responses of blood lead to differing sources are ideally estimated from regression analyses of epidemiological data. The form of this model is:

$$PbB = sC_s + aC_a + dC_d + wC_w + \ldots,$$

where PbB is the average blood lead concentration; s is the slope relating blood lead and soil lead; a, d, w, . . are slopes for air, diet, water . . ; and C_s, C_a, C_d, C_w, . . are lead concentrations in soil air, diet, water *etc*.

Proper use of this model would require measurement of all source terms as well as having good estimates for all the slopes.

The 'aggregate' model combines in a single slope or coefficient all the direct and indirect contributions of soil lead to blood lead, and also combines the contributions from all other sources into a single factor. It takes the form:

$$PbB = sC_s + B ,$$

where: s is the slope between blood lead and soil lead; C_s is the soil lead concentration; and B is the contribution to blood lead for all sources other than soil and dust. If any other source(s) make a significant contribution to blood lead, such as smelter emissions for example, this additional term must be measured or estimated and added to the background term B, in order to make a proper assessment of the contribution of soil lead to blood lead.

This approach has the advantage that it is easy to use and understand, and that the necessary data are generally available. Hence a modified version of the aggregate model was chosen for use in this report.

A more complex modelling approach is used in the 'biokinetic model'. Daily lead intake from indoor and outdoor air, food, water, dust and soil are calculated using age-specific estimates of parameters such as respiratory volume, amount of soil ingestion, and lead absorption for various routes of exposure through the lungs and gastrointestinal tract (Rabinowitz *et al.*, 1980). These estimates are then used to calculate an average blood lead concentration. Although this model relies on limited data for some input terms, it can be useful in predicting mean blood lead concentrations from multiple exposure sources and under alternate abatement strategies, because if the data are available, it allows for more adjustable parameters than the model adopted here.

The model used in this document is based upon a change in blood lead being equal to d times the soil lead concentration: change in PbB $= d\ C_s$. This can be related to the various factors used in the biokinetic model (USEPA, 1986,1988) and to the factors influencing uptake by the relationship:

$$\delta = (F.\ I.\ A.\ X\ ), \text{ where}$$

F = age-specific factor relating amount of absorbed lead (from any source) to blood lead concentration, taken as 0.4 for 2 year old children; I = soil or dust ingestion, mg d^{-1}; A = per cent absorption of dietary lead in the gut; X = product of factors effecting soil lead-blood lead relationship.

The multiple components of factor X can be represented as:

$$X = x_1.\ x_2.\ x_3.\ x_4.\ x_5.\ , \text{ where}$$

x_1 = bioavailability of soil or dust relative to normal dietary lead absorption (discussed in the bioavailability section of the report). x_2 = factor representing physical availability of soil/dust lead. Higher for dry, dusty sites, lower for grass covered sites; x_3 = factor indicating relative nutritional status of group. Nutritional deficiencies can enhance lead absorption; x_4 = a social/ethnic index. Well scrubbed, fully clothed children who spent all day indoors would be expected to have a lower soil uptake; x_5 = any other factors which may influence lead uptake from dust and soil.

As more data become available and if other criteria are established, other models rather than the one chosen here could of course be used within the framework established in this document.

5. Guidelines for Undeveloped Land.

In setting a guideline for publicly accessible land which is to be left undeveloped, or in which the lead in soil or dust is unavailable to any young children, say by completely grassing over the site, a modified approach should be adopted. The soil guideline derived by the model adopted here would be too restrictive on land usage and unnecessary to protect public health. Two possible approaches are suggested for such a situation.

(a). As soil and dust lead concentrations also follow log normal distributions, a level at two standard deviations (for example) above the guideline derived as above, could be used as a geometric mean guideline level for undeveloped sites. The appropriate formula would be:

$$U = S . G^n$$

where: U is the undeveloped land guideline, a geometric mean concentration. S is the guideline derived from the target blood lead concentration for developed land where children may be exposed to soil. G is the geometric standard deviation of the soil/dust lead concentrations, (typically about 2). n is the number of standard deviations chosen for the desired level of protection.

(b). Alternatively, a different 'level of protection' could be chosen in the original model formula. If, for example, n=3 were to be used for populated sites, such that 99.9% of the population should be below the target blood lead concentration, a lower value could be chosen for unpopulated areas, say n=0, equivalent to where the mean blood lead concentration would be below the target blood lead concentration.

7. Examples of Soil/Dust Guideline Calculations

Examples of how to calculate S, the soil or dust guideline value are given in Supplement 1, together with a computer program suitable for use on a personal computer, at the end of this report.

8. Lead in Soil/Dust Guideline Based on the Most Sensitive Individual

The guideline setting approach previously discussed is based upon deltas derived from population studies – in effect a mean response of blood lead to soil or dust lead. Rather than basing a guideline upon a certain percentage of the population being below a target blood lead concentration, say 98% less than 15 µg dl^{-1} for example, it may be desired to base a guideline on protecting the most sensitive individual. Such an individual would be the one experiencing adverse effects at the lowest PbB.

In its current biokinetic model, the USEPA (1988) makes use of the studies of Binder et al. (1986) and Clausing et al. (1987) on estimating the amount of soil ingested by the average child. Current estimates are that this is about 0.1 gd^{-1} for a two year old, with the 95th percentile about 0.5 gd^{-1} and the 99th percentile about 5 g d^{-1}. From the relationships previously discussed, delta may be taken to be proportional to the average daily amount of soil ingestion. Using the "best mean estimates" of 2 for delta and 0.1 g d^{-1} for soil/dust ingestion, a soil guideline can be calculated for various amounts of soil ingestion which would result in a maximum allowable increase in an individual's blood lead concentration above a baseline value.

Table 20. *Soil/dust guideline calculated for varying amounts of soil ingestion and baseline blood lead concentrations target PbB of 15 g dl⁻¹ assumed.*

Soil ingestion gd⁻¹	Soil/dust guideline, g Pb g⁻¹ Baseline blood lead, g dl⁻¹			
	0	2	4	6
0.1	7,500	6,500	5,500	4,500
0.5	1,500	1,300	1,100	900
1.0	750	650	550	450
2.5	300	260	220	180
5.0	150	130	110	90
10.0	75	65	55	45

Values for 0.1 g d⁻¹ ingestion calculated from $\{(T\text{-}B)/\delta\}$ 1,000, where delta is taken as 2. Values for other ingestion rates taken as directly proportional to those calculated for 0.1 g d⁻¹.

Table 20 illustrates how this may be done for a particular set of assumptions. Thus, taking a target PbB of 15 µg dl⁻¹, a baseline PbB of 4 µg dl⁻¹, and delta equal to 2, a guideline of 1,100 µg Pb/g would be set if one chose to protect 95% of children (0.5 g d⁻¹). Similarly, a guideline of 110 µg Pb/g would be used if one chose the 99th percentile of soil ingestion of 5 g d⁻¹ as a basis for protection of the most sensitive individual. Using these assumptions, a lead in soil guideline set on the basis of high soil ingestion by a child could easily result in a standard within the range of "normal" uncontaminated soil lead concentrations.

Values for 0.1 g/day ingestion calculated from $\{(T\text{-}B)/\delta\}$ −1000, where delta is taken as 2. As the calculations are for individuals, G and n are not applicable. Values for other ingestion rates are taken as directly proportional to those calculated for 0.1 g d⁻¹.

VIII. LEAD IN SOIL: RISK ASSESSMENT AND PROBLEM MANAGEMENT

A. Risk Assessment/Management

The overall analysis of risk due to an environmental contaminant has been systematised by the National Academy of Sciences (NRC, 1983). In this approach, the risk assessment is composed of two parts; namely, hazard assessment and exposure assessment. Associated with the risk assessment in the overall analysis is the area of risk management.

This Section is written primarily from a USA viewpoint and the examples chosen are American. Nonetheless, the underlying philosophy is widely applicable providing, in other nations, the relevant local environmental legislation is taken into account.

The main focus of the present discussion is on exposure assessment and, in particular, quantifying the relationship between blood lead levels and the levels of lead in soil as one possible source of exposure. The objective of this analysis is a *suggested soil lead guidance* based on the relationship between concentrations of lead in soil and the observed blood lead concentrations. This relationship forms part of the exposure assessment. Other parts of the exposure assessment include contributions to blood lead levels due to dust, water, food, paint and other sources. The overall exposure assessment includes all of these potential sources.

The specific blood lead content at which there is a health concern is a matter of present debate in the public health community as discussed in detail in Section VII of this report. This debate is leading to a better understanding of the problem and, hence, better agreement on the target or alert blood value. These revisions guide one in turn, to new evaluations of soil lead trigger content. Thus, this report proposes not setting *single* regulatory values for lead either in soil or blood but rather, establishing a soil value appropriate to the prevailing scientific knowledge and in the context of a systematic risk assessment.

The overall risk assessment can be achieved on a site specific or case-by-case basis. If the major exposure route is from the soil then the suggested guidance developed here can be applied directly to determine clean up levels.

Risk management has several inputs. One is the risk assessment as previously described. In this case the limited question of the relationship between blood lead levels and soil lead concentration is considered. However, it must be recognised that in deciding on clean-up strategies, several other factors impinge on the decision process. Other aspects of the risk assessment can change the decision such as the use of different blood lead levels of concern from the hazard assessment and exposures to lead from sources other than soil. Several other factors are involved in the risk management decision process; viz, economic, legal, political and social considerations. It is not the purpose of the current analysis to address these other aspects, only to note their existence.

The area of risk assessment has been considered by the Federal Government (NRC, 1983), Hallenbeck *et al* (1986), Ricci and Rowe (1985) and Rowe (1977) which may also serve as suggested reference material.

Risk Communication

There are three methodologies concerned with estimating risk: risk assessment, risk management and risk communication. Prior to 1986 there was little literature on the subject of risk communication. Since that time many articles have appeared and conferences and special sessions have been conducted on this topic. The importance of risk communication is in the increasing awareness of dissonances and tensions between the risk assessment experts and the lay communities. Risk communication represents a new policy focus and it considers the problem of the divergence between expert approaches and lay perceptions of risk. The broad question that underlies this subject is: how can experts and the general public communicate about uncertain environmental hazards in a manner that both educates the public, informs the experts and respects the democratic process? Risk communication has been discussed by Davies *et al.* (1987) and Covello *et al.* (1988).The National Academy of Sciences has provided guidabce on the process of risk assessment and how to differentiate fact from perception (NRC, 1983) and improve risk communication (NRC, 1989) A number of factors can contribute to the trust and confidence that can be established by a successful risk communication programme. These factors include; consistency in the risk estimate message and in the people communicating this message, independent corroboration of the risk assessments by external scientific advisory boards, easy public access to official regulatory agency information and data, understanding inconsistencies between scientific and popular views, not to assume that good risk communication correlates directly with a change in behaviour and collaboration between federal, state and local agencies.

A number of often incorrect and unspoken assumptions underlie the view some take to risk communication. Too often it is assumed that the risk assessment is done well and without bias and that such an assessment will lead all honourable experts to the same conclusions. Often it is assumed that the non-technical concerns of the lay public (fairness, local control, courtesy, property values and moral values to name a few) are irrelevant or of secondary concern and that therefore risk communication can be one-dimensional and technical instead of being (more realistically) multidimensional and value laden.

Uncertainties and Non-technical Considerations

Among the numerous technical and non-technical factors that need to be considered by the risk manager are the number and age of the exposed population. If the location of concern contains low income housing, school yards, or playgrounds, the issue is far more significant than if the area of concern contains factories, retirement communities or warehouses. The decision maker needs to consider present and probable future land use in deciding whether to act and if so what kind of remedial effort is required.

It is also important to recognise the probabilistic nature of this problem. While it is possible that 300 μg Pb g^{-1} in a soil may present a health problem to some children, it is more likely that 1,500 μg Pb g^{-1} will present a problem. Numerous factors such as the percentage of bare soil present; the number, age and ethnic group of the children; and the socio-economic status of the residents determine the extent of the problem. There are many areas that have soil leads greater than 1,000 μg Pb g^{-1}.It is important to begin cleaning up the largest number of areas containing the high risk children and the highest lead contamination. These are mostly the inner city areas where soil lead concentrations are high because of the past use of lead-based paint and leaded petrol as well as children from families in the lower income brackets.

While the bulk of this report concerns itself with scientific and technical issues, it is clear that many legal, political, social and economic aspects are important factors. Indeed, in many cases, these will be the deciding factors in favour of remedial actions.

There is always uncertainty in the scientific and technical data and models. This uncertainty will leave the decision maker (risk manager) without an absolute assessment of the risk. In many cases, the uncertainty will be so great that policy decisions regarding the appropriate margin of safety, the feasibility of remediation and the ultimate cost of clean-up will be the over-riding factors in the decision. In these cases risk communication will play an even more important part in the process than usual.

Geographic and Physical Processes that Affect Soil Lead Accumulation

An understanding of mechanisms involved in soil accumulations of lead are needed to clarify relationships between the various sources of lead and the responses of the population to these sources. The strong relationship between blood lead and soil lead has been well- described for a number of different populations living in a variety of socio-economic conditions. The most at risk populations show the strongest relationship, but the relationship is strong even for the least exposed middle to upper middle class children living in a suburban situation (Rabinowitz and Bellinger, 1988). There are several key features that assist with understanding the geographic distribution of lead in soil. These have been described in some detail in Section V.

The modern industrial city is composed of a complex combination of point sources and linear sources of lead which are unevenly spread over an area. The uneven distribution causes the lead content of urban soils to have a very distinctive pattern. Large cities have more accumulated lead than small towns of the same age. Within a given city, the amount of lead that has accumulated is usually highest in the neighbourhoods near the centre of the city and lowest in outlying suburbs. This pattern holds up even when the ages of the neighbourhoods are similar. Old neighbourhoods in outlying areas have significantly lower amounts of lead than similar aged neighbourhoods located towards the centre of the same city.

The basic patterns of soil lead in various urban environments match those described for the lead exposure of the population. The congruence between the soil lead and blood lead matches what has been learned about the relationship between soil lead and blood lead for children. The congruence between soil as an environmental measure and the exposure response by childhood populations provides planners and policy makers with an important tool for defining critical sites and for setting priorities for undertaking cost efficient amelioration of excessive lead exposure to children.

Uncertainties

It should be clear from the above discussion that the patterns of lead distribution are very complex within the urban environment. Although soil lead and blood lead are strongly related, the increase of blood lead per increase in the content of soil lead varies for different populations. The least exposed cleaner and well-maintained environments, experience a small increase per increment increase of lead in soil (less than 1 μg dl^{-1} /1,000 μg Pb g^{-1} soil for the least exposed upper middle class suburban children (Rabinowitz and Bellinger, 1988). The most exposed children exhibit a larger increase in blood lead per increment increase in soil lead. Several studies for the most exposed children have found levels of about 5–10 μg dl^{-1} /1,000 μg Pb g^{-1} soil (Angle and McIntyre 1982, Bornschein et al, 1986, Brunekrief et al., 1983). The greater sensitivity of the most exposed children is probably a function of the poor conditions of the environment (bare soils and play areas next to buildings), low nutritional status (especially low calcium and iron status), and perhaps an overall difficulty with supervision in a lead contaminated environment. It has been observed that the most exposed children have behavioural characteristics that make them hard to discipline. Their lead exposure predisposes them to lower quality of parental supervision (Dietrich et al., 1987).

The major uncertainty is how to proceed with amelioration. The above information provides some important guidance as to what can be done. Given the different rates of exposure in different places (e.g., the inner-city has a greater rate of exposure than suburban or rural areas) it should be clear that by focusing attention on those urban places that have the greatest content of lead will provide the largest amelioration benefit per unit expense of cleanup cost. Mapping the soil lead levels in the countryside and major cities (Davies et al., 1984, and Mielke and Adams 1989) can delimit those areas of greatest and immediate concern.

There are many uncertainties concerning how to proceed with lead prevention. For example, "deleading" homes once seemed like a good idea; but, because of the dust it generates, without thorough cleansing afterwards, it is not effective in preventing lead exposure to populations of young children. Demonstration programmes are needed to develop and test lead exposure prevention methods. There are several methods that should assist children. For example, if small particles are a major component of the problem, then the use of high efficiency vacuum cleaners (ULPA and HEPA) should reduce lead burdens from dust accumulations in homes of neighbourhoods with the highest lead content; improving the nutrition of the children who are most exposed should reduce their physiological response to lead. These approaches would be relatively inexpensive. Some of the most contaminated urban neighbourhoods may need outright soil removal and replacement in order to reduce the risk of lead exposure from soil and dust to acceptable levels. Other places may only need some resodding or grass seed in order to reduce lead levels to acceptable levels. Most non inner-city and small town neighbourhoods probably do not need any work at all.

Everyone should benefit from a carefully prepared and well focused educational programme about the lead problem. Herein lies a dilemma. The needs of a neighbourhood must be matched to the reality of the fact that there is not simply a single lead problem, rather there is a multi-dimensional lead problem. The problem is further confounded by the divergent cultural makeup of society and the complicated environmental lead patterns of the neighbourhoods of our modern urban society. Solving the lead exposure problem requires a number of approaches. It is important to be able to determine what methods are most effective as options for various types of neighbourhoods as lead prevention becomes part of public programmes.

Behavioural/Social Aspects of Lead Poisoning

The principal concern with respect to dust/soil lead is the young (6 - 72 months) child who inadvertently (or purposefully) ingests contaminated dust and/or soil (see Sections VI and VII). Clearly there are many social or economic or behavioural parameters that can increase the extent of exposure through hand-to-mouth activity in the child. These include the degree of cleanliness of the environment, the frequency of hand-to-mouth activity, the extent of parental supervision, the extent of contamination, and the extent to which absorption can be affected by factors such as nutritional status. While much remains to be understood about these interactions, there is a considerable amount of information available that gives insight on this complicated problem.

Socio/Economic Characteristics as Factors in Risk

A recent study by Pope (1986) estimated that the percentages of housing by year of construction having paint with lead greater than or equal to 0.7 mg cm^{-2} as: pre-1940, 99%; 1940–1959, 70%; and 1960–1975, 20%. Of particular interest are those homes containing lead-based paint which are not in good repair. Chisolm *et al.* (1985) and Clark *et al.* (1985) have shown that deterioration in housing has a very significant effect (as much as a doubling) on blood lead levels in young children. Clearly, deteriorated older housing stock will tend to be located in the poorer areas of the central cities.

The expectation of greater lead exposure among lower Socio-Economic Status (SES) children is supported by the NHANES II survey (Mahaffey *et al.*,1982) which found that the prevalence of blood lead levels greater than 30 µg dl^{-1} in 6 to 60 month old children in families with an annual income less than $6000 was nearly 10 times higher than in families with an annual income greater than $15,000. Other investigators such as Bornschein *et al.* (1985) have demonstrated the influence of educational background and degree of parental care on childhood blood lead levels.

It is quite likely that several factors contribute to the increased exposure of children from poor families to lead. Not only is there a higher incidence of dilapidated older housing, but there is also a tendency to find low income housing areas near busy central city streets (Mielke *et al.*, 1985) and there is a higher incidence of nutritional deficiencies that are associated with increased lead absorption by children (Yip *et al.*, 1981). In addition, factors such as an increased incidence of working single parents and accompanying poorer supervision because of poorer access to day care may contribute to the greater risk to children of low income families. Occupations of adults can also result in added exposure of children to lead. For example, contaminations of the home can occur from adults with lead dust on their clothing from lead related jobs.(Rice *et al.*, 1978)

Ethnic Grouping as a Risk Factor

The NHANES II data also exhibited a dramatic difference in lead exposure between white and black children in the range of 6 and 60 months of age with blood leads over 30 µg dl^{-1} . The rate of exposure of black children was found to be 2-4 times greater than white children. The situation for Hispanic children is less clear because the NHANES II data could not distinguish this group. A survey in New York found that Hispanic children were between white (Anglo) and black children in blood lead concentration (Billick *et al.*, 1979).

Because the NHANES II survey was not designed to distinguish Hispanics from Anglos and blacks, the Hispanic Health and Nutrition Examination Survey (HHANES) was conducted in 1982 to 1984 (Carter-Pokras *et al.*, 1990). This survey distinguished children of Mexican-American ancestry from those of Puerto Rican and Cuban ancestry. It also distinguished children of Mexican-American ancestry born in Mexico from those not born in Mexico. Unfortunately, because of the downward trend of blood lead levels (ATSDR, 1988) it is not possible to compare these groups with Anglos and blacks, but they can be compared with each other. In general, Puerto Rican children living in New York were at greater risk than the other sub populations. Mexican-American children born in Mexico were found to have higher blood lead levels than those born in the U.S.

Various social-economic indicators were also found to be significant in the HHANES study. Mean blood lead levels for Mexican-American and Puerto Rican children were highest for those children living in the central cities in families with the lowest annual incomes and education of head of household. Puerto Rican children living with a married head of household had lower mean blood lead levels than did children living with a single head of household.

It is possible that the ethnic background is simply a surrogate for a number of social/economic/cultural factors. The high proportion of Puerto Ricans who live in poverty (42%) may be the driving factor that contributes the exposure through a combination of intercity neighbourhoods, old dilapidated housing, poor nutrition, inadequate supervision (lack of day-care *etc.*) and other factors that can be attributed to poverty. In the same regard, Mexican-American children living in the Southwest have a 23% poverty rate compared to only 11% for all persons of non- Hispanic origin.

Thus, screening, investigation and remedial programmes should focus first on areas in the central cities with high populations of black and Puerto Rican children.

Age Distribution

Numerous studies have shown that urban children, particularly those of pre-school age (less than 5 years) are the sub-population most at risk (Mahaffey *et al,* 1982; Carter-Pokras *et al.,* 1990; ATSDR, 1988). As noted by the ATSDR report (1988), the precise age interval for children at greatest risk has not been defined. Because the exposure begins prenatally there is, in general, not a lower bound on the age. Since this report, concentrates on lead in dust and soil, it seems reasonable to assume that infants and toddlers will be at greatest risk from contaminated soils and dusts. The NHANES II study showed that 6 to 24 month old children had higher blood lead levels than 36 to 60 month children who in turn had higher blood lead levels than those older than 60 months.

Gender

The NHANES II study found that the percentage of elevated blood lead levels was slightly higher among boys than girls but this difference was not significant (Mahaffey *et al.,* 1982). In the HHANES study Mexican-American males had a statistically significantly higher mean blood lead level and a nonstatistically significantly higher percent elevated blood lead than did Mexican-American females (Carter-Pokras *et al.,* 1990).

Customs and Mores

In the Mexican-Hispanic and Hmong Cultures, folk remedies are often the cause of high blood lead levels and lead poisoning in children. *Azaran* and *greta* are fine powders with total lead contents varying from 70% to greater than 90%. They are often used by Mexican-Hispanics to treat children under 12 years of age for gastrointestinal illness. Hmong parents use a folk remedy referred to as "pay-loo-ah" to treat infants and children for rash and fevers. The remedy consists of red and orange powders with a lead concentration of 8%. Although there have been extensive educational programmes directed towards Hispanic and Hmong families publicising the dangers of these folk remedies, customs of use may still be retained in certain families and should be taken into consideration in any type of study or assessment (MMWR, 1983a, MMWR, 1983b).

Educational Background

Presumably, the educational background of concern is that of the parents. Because this is highly correlated with, and part of, social-economic status, there is an overlap with the discussion in the first paragraph of this section. The NHANES II results do not give any insight on the importance of parental education. The HISPANIC NHANES (Carter-Pokras *et al.*, 1990) data did show a significantly lower blood lead for Mexican-American children whose parents had a higher level of education. Since this factor interacts strongly with those involved in Social-Economics factors there does not seem to be a strong reason for considering it separately.

Legal Aspects Concerning Cleanup Levels for Lead

The EPA has not established a reference dose (RfD), or other level, which would set an acceptable daily intake for lead. As a result, the various EPA regions have experienced great difficulty in completing risk assessments for lead-contaminated sites. The absence of a national standard has resulted in inconsistent decisions on appropriate cleanup levels from the different EPA regions. This inconsistency has prompted the EPA's Office of Solid Waste and Emergency Response to advise the regions to use the Centers for Disease Control (CDC) guidance of 500 to 1,000 $\mu g\ g^{-1}$ for cleanup decisions (Inside EPA, Jan. 20, 1989, p. 15). Although there is as of yet no national policy, other lead levels have been established and may be considered during remedial activities.

The Occupational Safety and Health Act (OSHA) Air Contaminant Levels were designed to protect workers in closed environments such as buildings. OSHA has set an Action level at $30\mu g\ m^{-3}$ (averaged over an 8-hour period), and a Permissible Exposure Limit (PEL) of 50 $\mu g\ m^{-3}$ also averaged over an 8-hour period.

The Clean Air Act (CAA) has set a national primary and secondary ambient air quality standards for lead at $1.5\mu g\ m^{-3}$ with the maximum arithmetic mean averaged over a calendar quarter ("ambient air" is the portion of the atmosphere, external to buildings, to which the general public has access). The CAA allows for different emission limitations to be placed on each source of air pollution (see *e.g.*, 40CFR Part 60). The Clean Air Act also stipulates that individual states may adopt additional and more stringent limitations.

The Clean Water Act (CWA) has set National Primary Drinking Water Regulations (Maximum Contaminant Levels, MCLs) at 0.05 mg l^{-1} for lead. However, on August 18, 1988, the EPA proposed regulations which would place the MCL for lead at 0.005 mg^{-1}, and the MCL goal (MCLG) at zero. The CWA allows for different effluent limitations to be placed on each source of discharge into waters in the United States (see *e.g.* 40 CFR Part 433). The Clean Water Act also enables individual states to adopt additional and more stringent standards.

Potential Liability in Establishing Cleanup Levels

A small risk of liability exists if a person is injured at a "cleaned up" lead-contaminated site, and that injury can be traced to excessive lead concentrations remaining in the soil. However, the lawsuit potential can be reduced if a few precautions are taken. To begin with, the most likely basis of action is negligence, defined as "...the failure to observe, for the protection of the interests of another person, that degree of care, precaution, and vigilance which the circumstances justly demand, whereby such other person suffers injury." Therefore, the responsible party should make every attempt to assimilate all available studies and make an informed, scientific judgment based on those findings. This will insure that the responsible party is utilising "due care" in the determination. The investigation may find that "safe" lead levels will vary among contaminated sites, due to the differences in exposure potentials. This fact, and any other limiting data, should be included with the recommendations, so as to appropriately limit the scope of the findings.

A negligence cause of action would require a finding that the responsible party owed some duty to the suing party. This may be difficult to establish if: (1) no specific site is contemplated in establishing the guidelines, and (2) the recommendations emphasise that the necessary cleanup level may vary with each site, depending on potential exposures.

It should be noted that EPA and the States, under the current regulatory scheme, have the final authority to determine appropriate remedial levels. This determination – the Record of Decision (ROD) – must be made at each site. In addition, each ROD is published for public review and comment. By vesting final authority in the government, the process greatly reduces the likelihood that the responsible party would be held liable for excessive lead levels remaining after the remediation.

The most likely target of any lawsuit would be the government agency approving the cleanup level, and the parties who are responsible for the original contamination and/or subsequent cleanup of the site.

Economic Considerations in Establishing Lead Levels

After the remedial investigation is completed, a feasibility study is undertaken to develop and evaluate the remedial alternatives available at a particular site. Cost is considered at both the initial screening and detailed analysis stages of evaluation. Thus, during the initial screening of all alternatives developed in the feasibility study, the cost of implementing the remedial action must be considered which includes the operation and maintenance costs. An alternative which far exceeds the costs of other alternatives, without providing a substantially greater measure of protection to the public's health or the environment, nor increased technical reliability, should be excluded from further consideration. Those alternatives which meet or exceed the appropriate Federal public health and environmental requirements for "applicable or relevant and appropriate" clean up levels, are more desirable since they will provide greater protection than do those alternatives which do not meet such requirements.

After the initial screening, the remaining alternatives are evaluated in detail and the lead agency will select a "cost-effective remedial alternative" that properly mitigates and minimises threats to human health and the environment. In choosing the appropriate alternative, the lead agency will consider "cost, technology, reliability," and other concerns, with regards to their relevant effects. Also, the lead agency will typically consider costs only among those plans which meet the designated ARARs, although there are some statutory exemptions.

Cost-benefit analysis is made on a case-by-case basis. There is currently no set formula or ratio to apply in determining what degree of cleanup, due to excessive cost, would be viewed as inefficient.

EPA proposed revisions to the National Contingency Plan (NCP) on December 21, 1988. Cost effectiveness is still to be considered in selecting a remedy, but only after the alternatives have been found to provide adequate protection of human health and the environment. The selected alternative must also comply with all designated ARARs, or provide grounds for invoking a waiver of an ARAR.

Economic and Financial Considerations Concerning Remedial Actions

Before determining the need for a remedial action, it is necessary to establish a scope of the available financial resources. Depending on the site's location, size, and uses, the contaminated site may be eligible to enlist community, state and federal resources to supplement private funds, thus, ensuring that all necessary cleanup actions may be taken.

The costs involved with the physical cleanup of the soil are not always the only ones incurred during a remedial action. Liabilities involved in taking or not taking action also play a significant role in determining the scope of the remedial action plan. The costs of monitoring a site after the cleanup has been completed should also be factored into the total cost of the remedial action.

Remedial action methods which require some form of soil treatment to reduce potential health risks include, but are not exclusive to, the following:

Soil Removal

As the name suggests, this method involves excavating all of the lead contaminated soil and transporting it to an approved site for disposal. Clean fill is brought in, where it is necessary, to replace the soil excavated at the contaminated site.

This method is sufficient for most sites and serves as one of the quickest measures to immediately reduce the threat to public health. However, if the volume of contaminated soil requiring excavation is large, this method can be extremely costly. Another important point to remember is that the soil excavated from the site has not been decontaminated, but simply moved to another location and so it still contains unacceptable levels of the contaminating substance (ICRCL ,1987).

Soil Containment

Soil containment isolates the contaminated soil by covering it with new, clean soil or a hard cover. Hard cover is the preferred and more cost effective method of covering; however, to ensure that it works appropriately it must be well designed and properly installed and maintained (ICRCL, 1987).

A clean, inert fill can be used so long as it is sufficiently thick to contain the contaminated material and results in a soil which has an acceptable lead concentration. This process should not mix the old contaminated soil with the new covering soil and ground cover should be reintroduced as soon as possible to prevent new soil erosion (Elias, 1989).

Contaminant Extraction: Soil Washing and Flushing

These methods involve chemically treating the contaminated soil and, while they are effective at removing metals, they can be quite costly (ICRCL, 1987).

To wash contaminated soil, it is excavated and mixed on-site with a chemical capable of removing lead. The liquid is then extracted and the clean soil returned to the site. The chemical- lead extraction fluid can be treated and the lead removed for re- use at another site (Elias, 1989).

Soil flushing is a technique which can be used when a contaminant has already reached the ground water. The process is applied directly to the soil's surface. The solution is then given time to reach the ground water. Once that has occurred, the ground water is pumped up to the surface and treated. The clean ground water is then reapplied to the soil surface (Elias, 1989).

This method is cost-effective since it eliminates soil removal and replacement costs; however, it may require several applications before the contaminants are removed. Furthermore, soil flushing is not a closed, contained process and, therefore, runs the risk of further contaminating the ground water. Methods of modified soil flushing which are more controlled and contained have also been suggested and may prove useful in solving present limitations (Elias, 1989).

Deep Tilling

Instead of removing, covering or treating the contaminated soil, it can be tilled. Tilling mixes the contaminated soil with clean sub-soil, thereby reducing the lead levels on the surface (ICRCL, 1987). Tilling potentially reduces costs by eliminating the need for excavation and disposal; however, new costs may be incurred controlling drainage or erosion problems which may result (Elias, 1989).

Other Methods and Further Considerations

Revegetation of bare soils, community education, behaviour modification, zoning and vacuuming of dust within the house should all be considered as other methods of remedial action. The costs of these actions vary depending on the size of the population and the site itself. After remediation has been completed, the site must be monitored to ensure the cleanup action remains effective. The scale, duration, and cost of monitoring depends on the type of action taken. Finally, with time, the action should be assessed for future reference.

Costs of Not Doing Anything

The costs of undertaking a lead prevention project that is designed to reduce lead exposure are expensive Can society bear the cost of cleanup? Considering that lead paint removal and soil replacement costs thousands of dollars per home, the costs seem too much. This section considers some of the medical and remedial costs of lead exposure so that the price of not doing anything can be placed in perspective. The costs of excessive lead exposure are only partially quantifiable. For example, the costs to society of permanently limiting the potential of a child cannot be known. What can be known are some of the treatment and remedial costs associated with the most extreme forms of lead exposure.

An evaluation of some of the costs of lead poisoning were undertaken by Povenzano (1980), the USEPA (1985) and Szako and Pollack(1987) and Reagan and Mielke (1991). The last two documents provide estimated costs for the average medical and remedial education costs attributable to child lead poisoning (defined as 25 μgPb dL^{-1}) in Massachusetts and Minnesota. In Massachusetts, the 1986 estimated average medical cost was \$2,400 and estimated cost of remedial education was \$3,100 per lead poisoned child, and this results in about \$11 million spent for 2000 new lead poisoning cases which are found each year. In Minnesota, using a different model that included both **direct costs** from medical and educational expenses and **indirect costs** due to reduced future earnings, it was estimated that the lead poisoning of about 900 Minnesota children yearly resulted in a social cost of over \$12 million per year.

It is important to underscore what the above costs do not include. They do not include speech, physical and occupational therapy needed for many of these children. They do not include the extra costs in education (national average of about \$5,000 per pupil per year) for repeating grades in school that are often required by these children. The above costs do not include those incurred by a far larger group of children who may be experiencing adverse effects as a result of other levels of lead exposure. For example, exposure levels of 15m g dl-1 and even lower are now recognised as possibly being associated with learning and behavioural deficits and the costs of assisting this group of children are not included in the above costs. The costs incurred as a result of excessive lead exposure are so large that there are enormous benefits to society for expending efforts to prevent this problem.

SECTION IX REFERENCES ARRANGED BY SECTIONS.

SECTION II Introduction.

Davies, B.E., and Wixson, B.G. 1986. Lead in Soil: How Clean is Clean? In XX Annual Conference on Trace Substances in Environmental Health , Hemphill, D.D. (Editor.) Univ. of Missouri-Columbia 233-241.

Davies, B.E. and Wixson, B.G. (Editors). 1988. Lead in soil: issues and guidelines. Environmental Geochemistryand Health, Monograph Series No 4, supplement to Vol 9. Science Reviews Ltd., Northwood, U.K. 315 p.

Wixson, B.G. 1988. Overview on Lead in Soil: Issues and Guidelines Conference. In Lead in Soil: Issues and Guidelines . Davies, B.E. and Wixson, B.G. (Editors) Environmental Geochemistry and Health , Monograph Series 4, Supplement to Vol. 9. Science Reviews Ltd., Northwood, U.K., 1-6.

Wixson, B.G. 1989. Status Report on the Society for Environmental Geochemistry and Health Task Force on Lead in Soil. In XXIII Annual Conference on Trace Substances in Environmental Health . Hemphill, D.D. and Cothern, C. R. (Editors) Science Reviews Ltd., Northwood, U.K. 93 - 100.

SECTION III Glossary and definitions.

National Research Council. Committee on Pollution. 1966. Waste Management and Control National Academy of Sciences, Washington, DC (NAS Publication 1400) 257 pp.

Simms, D.L. and Beckett, J.J. 1986. Contaminated Land: Setting trigger concentrations. The Science of the Total Environment , 65 , 121-134

Wanielista, M.P., Yousef, Y.A., Taylor, J.S. and Cooper, C.D. 1984. Engineering and the Environment . Brooks/Cole Engineering Div. Wadsworth, Inc. 3.

Warren, C.E. 1971. Biology and Water Pollution Control . W. B. Saunders Co. 11-14.

SECTION IV Phased action plan.

U.S. Environmental Protection Agency. May 1979. Quality Assurance Requirements for State and Local Air Monitoring Stations (SLAMS) . Federal Register,Washington D.C. Vol. 44, No. 92, 27574-81.

SECTION V Lead in soil.

Ahrens L.H. 1954. The lognormal distribution of the elements. Geochimica Et Cosmochimica Acta, 5, 49 - 73.

Ahrens L.H. 1966. Element distributions in specific igneous rocks. Geochimica Et Cosmochimica Acta, 30, 109 - 122.

Alloway B.J. and Davies B.E. 1971. Trace element content of soils affected by base metal mining in Wales. Geoderma, 5, 197 - 208.

Berrow M.L. and Webber J. (1972) Trace elements in sewage sludges. Journal of Science of Food and Agriculture, 23, 93 - 100.

Bolviken B. and Lag J. (1977) Natural heavy-metal poisoning of soils and vegetation: an exploration tool in glaciated terrain Transactions Institute Mining and Metallurgy, 86, B173-B180.

Cannon H.L. and Bowles J.M. (1962) Contamination of vegetation by tetraethyl lead. Science, 137, 765 - 766.

Coker E.G. and Matthews P.J. (1983) Metals in sewage sludge and their potential effects in agriculture. Water Science Technology, 15, 209 - 225.

Culbard E.B., Thornton I., Watt J., Wheatley M., Moorcroft S. and Thompson M. (1988) Metal Contamination in British Urban Dusts and Soils Journal of Environmental Quality, 17, 226-234.

Davies B.E. (1978) Plant-available lead and other metals in British Garden Soils Science of the Total Environment, 9, 243-262.

Davies B.E. (1983) A graphical estimation of the normal lead content of some British soils. Geoderma, 29, 67 - 75.

Davies B.E. and Ballinger R.C. (1990) Heavy metals in North Somerset, England with special reference to contamination from base metal mining in the Mendips. Environmental Geochemistry and Health, 12, 291 - 300.

Davies B.E., Conway D. and Holt S. (1979) Lead pollution of London soils: a potential restriction of their use for growing vegetables Journal of Agricultural Science (Cambridge), 93, 749-752.

Davies B.E. and Houghton N.J. (1983) The selenium content of Welsh soils with special reference to bedrock and contamination from sulphide ores Minerals and the Environment, 5, 67-70.

Davies B.E. and Houghton N.J. (1984) Distance-decline patterns in heavy metal contamination of soils and plants in Birmingham England. Urban Ecology, 8, 285 - 294.

Davies B.E. and Roberts L.J. (1975) Heavy metals in soils and radish in a mineralised area of Wales, Great Britain. Science of the Total Environment, 4, 249 - 261.

Davies B.E. and Roberts L.J. (1978) The distribution of heavy metal contaminated soils in north east Clwyd, Wales. Water, Air and Soil Pollution, 9, 507 - 518.

Davies B.E. and Wixson B.G. (1985) Trace elements in surface soils from the mineralised area of Madison county, Missouri, U.S.A. Journal of Soil Science, 36, 551 - 570.

Frank R., Ishida K. and Suda P. (1976) Metals in agricultural soils of Ontario. Canadian Journal of Soil Science, 56, 181 - 196.

Griffith J.J. (1919) Influence of mines upon land and livestock in Cardiganshire. Journal of Agricultural Science (Cambridge), 9, 366 - 395.

Holmgren G.G.S., Meyer M.W., Daniels R.B., Kubota J. and Chaney R.L. (1993) Cadmium, lead, zinc, copper and nickel in agricultural soils of the United States. Agronomy Abstracts, 1983, 33.

Lagerwerff, J.V., Brower, D.L. and Biersdorf, G.T. (1973). Accumulation of cadmium, copper, lead and zinc in soil and vegetation in the proximity of a smelter. In VI Annual Conference on Trace Substances in Environmental Health, Hemphill, D.D. (Ed.) University of Missouri, Columbia. 71-78

Le Riche H.H. (1968) Metal contamination of soil in the Woburn market 1garden experiment resulting from the application of sewage sludge. Journal of Agricultural Science (Cambridge), 71, 205 - 208.

Merry R.H., Tiller K.G. and Alston A.M. (1983) Accumulations of copper, lead and arsenic in some Australian orchard soils. Australian Journal of Soil Research, 21, 549 - 561.

Mielke, H.W. (1993). Lead Dust Contaminated U.S.A. Communities: Comparison of Louisiana and Minnesota. Applied Geochemistry 6: 1-5.

Mielke, H.W. and Adams, J.L. (1989). Environmental Lead Risk in the Twin Cities. Center for Urban and Regional Affairs. Minneapolis, MN.

Mielke, H.W., Adams, J., Reagan, P.L., and Mielke P.W.Jr. 1988. Soil-dust lead and childhood lead exposure as a function of city size and community traffic flow: The case for lead abatement in Minnesota. In Davies and Wixson (Editors) Lead in soil: issues and guidelines. Environmental Geochemistry and Health , Monograph Series No 4, supplement to Vol 9. Science Reviews Ltd. Northwood, U.K., 253-271.

Mielke, H.W., Anderson, J.C., Berry, K.J., Mielke, P.W., Chaney, R.L. and Leech. M. (1983). Lead concentrations in inner-city soils as a factor in the child lead problem. American Journal of Public Health, 73, 1366 - 1369.

Mielke, H.W., Burroughs, S., Wade, R., Yarrow, T. and Mielke, P.W. Jr. 1984/85. Urban lead in Minnesota: soil transect results of four cities. Journal of the Minnesota Academy of Science 50 (1): 19-24.

Nriagu, J.O. 1978. Lead in soils, sediments and terrestrial rocks. (in) The Biogeochemistry of Lead in the Environment, Nriagu,J. (Editor) Elsevier/North HollandBiomedical Press, Amsterdam

Pierce F.J., Dowdy R.H. and Grigal D.F. (1982) Concentrations of six trace metals in some major Minnesota soil series. Journal of Environmental Quality, 11, 416 - 422.

Purves D. (1966) Contamination of urban garden soils with copper and boron. Nature (London), 210, 1077 - 1078.

Ragaini R.C., Ralston H.R. and Roberts N. (1977) Environmental trace contamination in Kellogg, Idaho, near a lead smelting complex. Environmental Science and Technology, 11, 773 - 781.

Reaves, G.A. and Berrow, M.L. 1984. Total lead concentrations in Scottish soils. Geoderma, 32, 1 - 8.

Rose, A.W., Hawkes, H.E. and Webb, J.S. (1979) Geochemistry in Mineral Exploration; Academic Press, London: 2nd edition.

Sinclair A.J. (1974) Selection of thresholds in geochemical data using probability graphs. Journal of Geochemical Exploration, 3, 129 - 149.

Smith W.H. (1976) Lead contamination of the Roadside Ecosystem Journal of Air Pollution Control Association, 26, 753-766.

Sommers, L.E., Nelson, D.W. and Yost, K.J. (1976). Variable nature of chemical composition of sewage sludges. Journal of Environmental Quality, 5, 303 - 306.

Tennant C.B. and White M.L. (1959) Study of the distribution of some geochemical data. Economic Geology, 54, 1281 - 1290.

Warren H.V. and Delavault R.E. (1960) Observations on the biogeochemistry of lead in Canada. Transactions of the Royal Society of Canada, 54, 11 - 20.

Wheeler G.L. and Rolfe G.L. (1979) The Relationship between daily traffic volume and the distribution of lead in roadside soil and vegetation Environmental Pollution, 18, 265-274.

SECTION VI Biovailability of lead for animals.

Allcroft, R. 1950. Lead as a nutritional hazard to farm livestock. IV. Distribution of lead in the tissues of bovines after ingestion of various lead compounds. Journal of Comparative Pathology 60:190-208.

Angle, C.R., Stelmak, K.L. and McIntire, M.S. (1975). Lead and iron deficiency. In IX Annual Conference on Trace Substances in Environmental Health. Hemphill, D.D. (Ed.) University of Missouri, Columbia, 377 - 386.

Barltrop, D. and Meek, F. 1975. Absorption of different lead compounds. Postgraduate Medical J. 51:805-809.

Barltrop, D. and Meek, F. 1979. Effect of particle size on lead absorption from the gut. Archives Environmental Health 34:280-285.

Barltrop, D., Strehlow, C. D., Thornton, I. & Webb, J.S. 1975. Absorption of lead from dust and soil. Postgraduate Medical J. 51, 801-804.

Bates, G.W. 1990. Personal Communications. Texas A. M University. College Station, Texas.

Bell, R.R. and Spickett, J.T. 1981. The influence of milk in the diet on the toxicity of orally ingested lead in rats. Food Cosmet. Toxicol. 19:429-436.

Bezwoda, W., R. Charlton, T. Bothwell, J. Torrance, and F. Mayet. 1978. The importance of gastric hydrochloric acid in the absorption of nonheme food iron. J. Lab. Clin. Med. 92:108-116.

Blake, K.C.H., Barbezat, G.O. and Mann, M. 1983. Effect of dietary constituents on the gastrointestinal absorption of [203]Pb in man. Environ. Res. 30:182-187.

Blake, K.C.H., and Mann, M. 1983. Effect of calcium and phosphorus on the gastrointestina absorption of [203]Pb in man. Environ. Res. 30:188-194.

Calabrese, E.J., Barnes, R., Stanek, E.J., Pastides, H.,Gilbert, C.E., Veneman, P., Wang, X., Lasztity, A. and Kostecki, P.T. 1989. How much soil do young children ingest: an epidemiological study. Register Toxicology and Pharmacology, 10, 73 - 82.

Carlson, A.J. and Woefel, A. 1913. The solubility of white lead in human gastric juice, and its bearing on the hygiene of the lead industries. Am. J. Public Health 3:755-769.

Chamberlain, A.C. 1987. Tracer experiments with lead isotopes. In: Thornton, I. and Culbard, E. (Eds.) Lead in the Home Environment: Sources, Transfer and Exposure Assessment. Science Reviews Ltd., Northwood, England. 179-188.

Chamberlain, A.C., Heard,M.J., Little,P., Newton,D., Wells, A.C. and Wiffen, R.C. 1978. Investigations Into Lead From Motor Vehicles. A.E.R.E. Report R-9188. H.M.S.O., London.

Chaney, R.L., Sterrett, S.B. and Mielke, H.W. 1984. The potential for heavy metal exposure from urban gardens and soils. pp. In Proc. Symp. Heavy Metals in Urban Gardens. J.R. Preer (ed.) Agric. Exp. Sta., Univ. Dist. Columbia, Washington. 37-84.

Chaney, R.L., Mielke, H.W., and Sterrett, S.B. 1989. Speciation, mobility and bioavailability of soil lead. In Lead in Soil: Issues and Guidelines . Davies, B.E. and Wixson, B.G. (Editors) Environmental Geochemistry and Health , Monograph Series 4, Supplement to Vol. 9. Science Reviews Ltd., Northwood, U.K. 105-130.

Dacre, J.C., and Ter Haar, G.L. 1977. Lead levels in tissues from rats fed soils containing lead. Archives Environmental Contamination Toxicology. 6:111-119.

Davis, S,, Waller, P., Buschbom, R., Ballou, J. and White, P. 1990. Quantitative estimates of soil ingestion in normal children between the ages of 2 and 7 years: population-bases estimates using aluminum, silicon and titanium as soil tracer elements. Archives Environmental Health, 45, 112 - 122.

Day, J.P., Ferguson, J.E. and Chee, T.M. 1979. Solubility and potential toxicity of lead in urban street dust. Bull. Environ. Contam. Toxicol. 23:497-502.

Decker, A.M., Chaney, R.L., Davidson, J.P., Rumsey, T.S., Mohanty, S.B. and

Hammond, R.C. 1980. Animal performance on pastures topdressed with liquid sewage sludge and sludge compost. In Proc. Nat. Conf. Municipal and Industrial Sludge Utilization and Disposal. Information Transfer, Inc., Silver Spring, MD. 37-41.

Duggan, C.J. and Inskip, M.J. 1985. Childhood exposure to lead in surface dust and soil: a community health problem. Public Health Rev. 13: 1-54.

Egan, D.A., and O'Cuill, T. 1970. Cumulative lead poisoning in horses in a mining area contaminated with Galena. Vet. Rec. 86:736-738.

Elfving, D.C., Haschek, W.M., Stehn, R.A., Bache, C.A. and Lisk, D.J. 1978. Heavy metal residues in plants cultivated on and in small mammals indigenous to old orchard soils. Arch. Environ. Health 33:95-99.

Ferguson, J.E., Forbes, E.A., Schroeder, R.J. and Ryan, D.E. 1986. The elemental composition and sources of house dust and street dust. Sci. Total Environ. 50:217-221.

Flanagan, P.R., Chamberlain, M.J. and Valberg, L.S. 1982. The relationship between iron and lead absorption in humans. Am. J. Clin. Nutr. 36:823-829.

Fries, G.F., Marrow, G.S., and Snow, P.A. 1982. Soil ingestion by swine as a route of contaminant exposure. Environ. Toxicol. Chem. 1:201-204.

Gibson, M.J., and Farmer, J.G. 1984. Chemical partitioning of trace element contaminants in urban street dirt. Sci. Total Environ. 33:49-57.

Hammond, R.C. 1980. Animal performance on pastures topdressed with liquid sewage sludge and sludge compost. In Proc. Nat. Conf. Municipal and Industrial Sludge Utilization and Disposal. Information Transfer, Inc., Silver Spring, MD. 37-41

Harbourne, J.F., McCrea, C.T. and Watkinson, J. 1968. An unusual outbreak of lead poisoning in calves. Vet. Rec. 83:515-517.

Harrison, R.M., Laxen, D.P.H. and Wilson, S.J. 1981. Chemical associations of lead, cadmium, copper, and zinc in street dusts and roadside soils. Environ. Sci. Technol. 15:1378-1383.

Healy, M.A. 1984. Theoretical model of gastrointestinal absorption of lead. J. Clin. Hosp. Pharm. 9:257-262.

Healy, M.A., Harrison, P.G., Aslam, M., Davis, S.S., and Wilson, C.G. 1982. Lead sulphide and traditional preparations: Routes for ingestion, and solubility and reactions in gastric fluid. J. Clin. Hosp. Pharm. 7:169-173.

Heard, M.J., Chamberlain, A.C. and Sherlock, J.C. 1983. Uptake of lead by humans and effect of minerals and food. Sci. Total Environ. 30:245-253.

Heyworth, F., Spickett, J., Dick, M., Margetts, B., and Armstrong, B. 1981. Tailings from a lead mine and lead levels in school children: a preliminary report. Med. J. Australia. 2, 232-234.

Hutton, M., and Goodman, G.T. 1980. Metal contamination of feral pigeons

Columbia livia from the London area: 1. Tissue accumulation of lead, cadmium, and zinc. Environ. Pollut. A22:207-217.

Ireland, M.P. 1977. Lead retention in toads (Xenopus laevis) fed increasing levels of lead-contaminated earthworms. Environ. Pollut. 12:85-92.

James, H.M., Hilburn, M.E. and Blair, J.A. 1985. Effects of meals and meal times on uptake of lead from the gastrointestinal tract in humans. Human Toxicol. 4:401-407.

Johnson, D.E., Kienholz, E.W., Baxter, J.C., Spanger, E. and Ward, G.M. 1981. Heavy metal retention in tissues of cattle fed high cadmium sewage sludge. J. Anim. Sci. 52:108-114.

Johnson, L.R. 1990. Personal Communications. University of Tennessee, Memphis, Tennessee.

Kienholz, E., Ward, G.M., Johnson, D.E., Baxter, J.C., Braude, G.L. and Stern, G. 1979. Metropolitan Denver sewage sludge fed to feedlot steers. J. Anim. Sci. 48:735-741.

Longstreth, G.F., Malagelada, J.R., and Go, V.L.W. 1975. The gastric response to a transpyloric duodenal tube. Gut 16:777-780.

Lucey M.R., Wass,J.A.H., Rees,L.H., Dawson,A.M., and Fairclough, P.D. 1989. Relationship between gastric acid and elevated plasma somatostatinlike immunoreactivity after a mixed meal. Gastroenterol. 97:867-872.

Mahaffey, K.R., 1981. Nutritional factors in lead poisoning Nutrition Reviews , 39, 353-362.

Mahaffey, K.R. 1985. Factors modifying susceptibility to lead toxicity. In Dietary and Environmental Lead: Human Health Effects. Mahaffey, K.R. (Ed.) Elsevier Biomedical Press, Amsterdam.373-419.

Mahaffey, K.R., Banks, T.A., Stone, C.L., Capar, S.G., Compton, J.F. and Gubkin, M.H. 1977. Effect of varying levels of dietary calcium on susceptibility to lead toxicity. Proc. Intern. Conf. Heavy Metals in the Environ. 3:155-164.

Mahaffey, K.R. and Michaelson, I.S. 1980. The interaction between lead and nutrition. In: Low Level Lead Exposure: The Clinical Implications of Current Research. Needleman, H.L. (Ed.) Raven Press, Boston. 159-200.

Malagelada, J.-R., Go,V.L.W., and Summerskill,W.H.J. 1979. Different gastric, pancreatic, and biliary responses to solid-liquid or homogenized meals. Dig. Dis. Sci. 24:101-110.

Malagelada,J.R., Longstreth,G.F., Summerskill,W.H.J. and Go, V.L.W. 1976. Measurement of gastric functions during digestion of ordinary solid meals in man. Gastroenterol. 70:203-210.

Malagelada, J.-R., Longstreth,G.F., Deering,T.B., Summerskill,W.H.J., and Go, V.L.W. 1977. Gastric secretion and emptying after ordinary meals in duodenal ulcer. Gastroenterol. 73:989-994.

Middaugh, J.P., Li,C., and Jenkerson, S.A. 1989. Health hazard and risk assessment from exposure to heavy metals in ore in Skagway, Alaska. Final Report, Oct. 23, 1989. State of Alaska Dept. Health and Social Services.

Mielke, H.W., and Heneghan,J.B. 1991. Selected chemical and physical properties of soils and gut physiological processes that influence lead bioavailability. Chemical Speciation and Bioavailability 3: 129-134.

Miller, D.D., Schricker,B.R., Rasmussen,R.R. and Van Campen,D. 1981. An in vitro method for estimation of iron availability from meals. Am. J. Clin. Nutr. 34:2248-2256.

Mylroie, A.A., Moore, L., Olyai, B. and Anderson, M. 1978. Increased susceptibility to lead toxicity in rats fed semipurified diets. Environ. Res. 15:57-64.

Nelson, L.S., Jr., Jacobs, F.A. and Brushmiller, J.G. 1985. Solubility of calcium and zinc in model solutions based on bovine and human milks. J. Inorg. Biochem. 24:255-265.

Nelson, L.S., Jr., Jacobs, F.A. and Brushmiller, J.G. 1987. Coprecipitation modulates the solubility of minerals in bovine milk. J. Inorg. Biochem. 19:173-179.

Paskins-Hurlburt, A.J., Tanaka, Y., Skoryna, S.C., Moore, W., Jr. and Stara, J.F. 1977. The binding of lead by a pectic polyelectrolyte. Environ. Res. 14:128-140.

Peaslee, M.H. and Einhellig, F.A. 1977. Protective effect of tannic acid in mice receiving dietary lead. Experientia 33:1206.

Rabinowitz, M.B., Kopple, J.D. and Wetherill, G.W. 1980. Effect of food intake and fasting on gastrointestinal lead absorption in humans. Am. J. Clin. Nutr. 33:1784-1788.

Reddy, M.B., Browder,E.J. and Bates,G.W. 1988. Cannulated swine and in vitro approaches to iron bioavailability. In Essential and Toxic Trace Elements in Human Health and Disease. Prasad, A.S. (Ed.). A.R. Liss, Inc., New York. 173-185.

Roy, B.R. 1977. Effects of particle sizes and solubilities of lead sulphide dust on mill workers. Am. Ind. Hyg. Assoc. J. 38:327-332.

Scanlon, P.F., Kendall, R.J., Lochmiller, R.L. and Kirkpatrick, R.L. 1983. Lead concentrations in pine voles from two Virginia orchards. Environ. Pollut. B6:157-160.

Schricker, B.R., Miller,D.D., Rasmussen,R.R. and Van Campen,D. 1981. A comparison of in vivo and in vitro methods for determining availability of iron from meals. Am. J. Clin. Nutr. 34:2257-2263.

Stara, J., Moore, W., Richards, M., Barkely, N., Neiheisel, S. and Bridbord, K. 1973. Environmentally bound lead. III. Effects of source on blood and tissue levels in rats. In EPA Environmental Health Effects Research Series A-670/1-73-036. Washington D.C. 28-29.

Steele, M.J., Beck,B.D., Murphy,B.L. and Strauss,H.S. 1990. Assessing the contribution from lead in mining wastes to blood lead. Regulat. Toxicol. Pharmacol. 11: 158-190.

Stone,C.L., Fox, M.R.S., and Hogye, K.S. 1981. Bioavailability of lead in oysters fed to young Japanese quail. Environ. Res. 26: 409-421.

Takeuchi,K., Peitsch,W., and Johnson,L.R. 1981. Mucosal gastrin receptor. V. Development in newborn rats. Am. J. Physiol. 240:G163-G169.

Thornton, I. 1986. Lead in United Kingdom soils and dusts in relation to environmental standards and guidelines. In XX Annual Conference on Trace Substances in Environmental Health. Hemphill, D.D. (Ed.) University of Missouri, Columbia. 298-307.

USCDC. 1985. Preventing lead poisoning in young children: a statement by the Centres for Disease Control - January 1985. US DHHS No. 99-2230. Atlanta, Georgia. 35pp.

USEPA. 1986. Air quality criteria for lead. Research Triangle Park, NC. Environmental Criteria and Assessment Office. 4 vols. EPA-600/8-83/028cF. Washington D.C.

Van Wijnen, J.H., Clausing P. and Brunekreef, B. 1990. Estimated soil ingestion by children. Environmental Research, 51, 147 - 162.

Wardrope, D.D. and Graham, J. 1982. Lead mine waste: hazards to livestock. Vet. Rec. 111:457-459.

Watson, W.S., Hume, R. and Moore, M.R. 1980. Oral absorption of lead and iron. Lancet 2:236-237.

Wise, A. 1981. Protective action of calcium phytate against acute lead toxicity in mice. Bull. Environ. Contam. Toxicol. 27:630-633.

Woefel, A., and Carlson,A.J. 1914. The solubility of lead sulphide ores and of lead sulphide in human gastric juice. J. Pharmacol. Exper. Ther. 5:549-552.

Young, R.W., Ridgely, S.L., Blue, J.T., Bache, C.A. and Lisk, D.J. 1986. Lead in tissues of woodchucks fed crown vetch growing adjacent to a highway. J. Toxicol. Environ. Health 19:91-96.

SECTION VII Lead in human health.

Angle, C.R. & McIntire, M.S. 1982. Children, the barometer of environmental lead. Adv. Pediatr. 27, 3-31.

Angle, C.R. & McIntire, M.S. 1979. Environmental lead and children: the Omaha study. J. Toxicol Environ Health 5, 855-870.

Annest, J.L., Pirkle, J.L., Makuc, D., Neese, J.W., Bayse, D.D. and Kovar, M.G.. 1983. Chronological trend in blood lead levels between 1976 and 1980. N Engl. J. Med. 308 , 1373-1377.

ATSDR (Agency for Toxic Substances and Disease Registry). 1988. The Nature and Extent of Lead Poisoning in Children in the United States: A report to Congress. U.S. Department of Health and Human Services. Atlanta, Georgia.

Baghurst, P.A., McMichael, A.J., Wigg, N.R., Vimpani, G.V. Robertson, E.F., Roberts R.J. and Tong, S.C. 1992. Environmental exposure to lead and childrens intelligence at the age of seven years: The Port Pirie cohort study. N. Engl. J. Med. 327: (18) 1279-1284.

Baker, E.L., Folland, D.S., Taylor, T.A., Frank, M., Peterson, W., Lovejoy, G., Cox, D., Houseworth, G. & Landrigan, P.J. 1977. Lead poisoning in children of lead workers. N Engl J Med 296, 260-261.

Barltrop, D. 1966. The prevalence of pica. Am J. Dis. Child. 112 , 116-123.

Barltrop, D. 1969. Transfer of lead to the human fetus. In: Mineral metabolism in pediatrics. Barltrop, D., Burland, W.L., (Eds.) Philadelphia, PA: Davis Co., 135-151.

Barltrop, D. Strehlow, C.D. Thornton, I. & Webb, J.S. 1975. Absorption of lead from dust and soil. Postgraduate Medical Journal 51, 801 - 804.

Barry, P.I. 1975. A comparison of concentrations of lead in human tissues. Br. J. Ind. Med. 32 , 119-139.

Bellinger, D., Leviton, A., Waternaux, C., Needleman, H.L. and Rabinowitz, M.B. 1987. Longitudinal analysis of prenatal and postnatal lead exposure and early cognitive development. N. Engl. J. Med. 316 , 1037-1043.

Bellinger, D., Leviton, A. and Sloman, J. 1990. Antecedents and correlates of improved cognitive performance in children exposed *in utero* to low levels of lead. Environmental Health Perspectives, 89, 5 - 11.

Bellinger, D., Slowman, J., Leviton, A., Rabinowitz, M., Needleman, H.L. and Waternaux. C. 1991. Low level lead exposure on children's cognitive function in the pre-school years. Pediatrics, 87, 219 - 227.

Bellinger, D., Stiles, K.M. and Needleman, H.L. 1992. Low level lead exposure, intelligence and academic achievement: a long term follow up study. Pediatrics 90: 855-861.

Byers, R.K. and Lord, E. 1943. Late effects of lead poisoning on mental development. AJDC 66:471-494.

Binder, S., Sokal, D. & Maugham, D. 1986. Estimating soil ingestion: the use of tracer elements in estimating the amount of soil ingested by young children. Arch. Env. Health 41, 341-345.

Boggess, W.R. and Wixson, B.G. 1977. (Editors). Lead in the Environment. National Science Foundation, NSF/RA-770214 Washington, DC. 272 pages.

Bornschein, R.L., Clark, C.S., Grote, J., Road, S. & Succop, P. 1989, Soil lead-blood lead relationship in a former lead mining town. In Lead in Soil: Issues and Guidelines. Davies, B.E. and Wixson B.G. (Editors) Environmental Geochemistry and Health. Monograph Series 4, Supplement to Vol 9. Science Reviews Ltd. Northwood, England. 149-160.

Bornschein, R.L., Succop, P.A., Krafft, K.M., Clark, C.S., Peace,B. and Hammond, P.B. 1986. Exterior surface dust lead, interior house dust lead

and childhood lead exposure in an urban environment. In: XX Annual
Conference of Trace Substances in Environmental Health. Hemphill,
D.D., (Ed.). University of Missouri, Columbia. 322-332.

Byers, R.K. and Lord, E. 1943. Late effects of lead poisoning on mental
development. AJDC 66:471-494.

US CDC. 1985. Preventing lead poisoning in young children: A statement by
the Centers for Disease Control, January 1985. US-DHHS No. 99-2230.
Atlanta, Georgia. 35 pp.

US CDC. 1991. Preventing lead poisoning in young children: a statement by
the Centers for Disease Control, October 1991. Atlanta, Georgia.

Chisolm, J.J., Jr. and Barltrop, D. 1979. Recognition and management of
children with increased lead absorption. Arch. Dis. Child., 54 , 249-262.

Chisolm, J.J., Jr. 1985. Ancient sources of lead and lead poisoning in the
United States today West J. Med., 143 , 380-381.

Chisolm, J.J., Jr. 1978. Fouling one's own nest. Pediatrics, 62 , 614-617.
Chisolm, J.J., Jr., Thomas, D.J. and Hamill, T. 1985. Erythrocyte
porphobilinogen synthase activity as an indicator of lead exposure in
children. Clin. Chem. 31, 601-605.

Clausing, P., Brunkreef, B. and van Wijnen, J.H. 1987. A method for
estimating soil ingestion by children. Int. Arch. Occup. Environ. Health
59, 73-82.

Cooney, G.H., Bell, A. McBride, W., and Carter, C. 1989. Low-level exposures
to lead: The Sydney lead study. Devel. Med. Child Neur. 31: 640-649.
Cooney, G.H., Bell, A., McBride, W., and Carter, C. 1988.
Neurobehavioural consequences of prenatal low level exposures to lead.
Neurotox. Terat. 11:95-114.

Cory-Slechta, D.A., Weiss, B., and Cox, C. 1987. Mobilization and
re-distribution of lead over the course of calcium disodium ethylene
diamine tetraacetate chelation therapy. J. Pharmacol Exp. Ther. 243,
804-813.

Davis, J.M. and Svendsgaard, D.J. 1987. Lead and child development.
Nature, 329, 297-300.

Dietrich, K.N., Berger, O.G., Succop, P.A., Hammond, P.B. and Bornschein,
R.L. 1993(a) The developmental consequences of low to moderate
prenatal and post natal lead exposure: Intellectual attainment in the
Cincinatti lead study cohort following school entry. Neurotoxicology and
Teratology 15: in press.

Dietrich, K.N., Berger, O.G., and Succop, P.A. 1993(b) Lead exposure and the
motor developmental status of urban six year old children in the
Cincinatti prospective study. Pediatrics 91: in press.

Dietrich, K.N., Succop, P.A., Bornschein, R.L., Krafft, K.M., Berger, O.
Hammond, P.B., and Buncher, C.R. 1990. Lead exposure and

neurobehavioral development in later infancy. Environ. Health Persp. 89:13-19.

Dietrich, K.N., Krafft, K.M., Bier, M., Succop, P.A., Berger, O., and Bornschein, R.L. 1986. Early effects of fetal lead exposure: neurobehavioral findings at 6 months. Int. J. Biosocial Res. 8: 151-168.

Dietrich, K.N., Krafft, K.M., Pearson, D.J., Harris, L.C., Bornschein, R.L., Hammond, P.B., and Succop, P.A. 1985. Contribution of social and developmental factors to lead exposure during the first years of life. Pediatrics 75:1114-1119.

Duggan, M.J. and Inskip, M.J. 1985. Childhood exposure to lead in surface dust and soil: a community health problem. Public Health Rev. 13: 1-54. Pediatrics 67: 911-919.

Duggan, M.J., Inskip, M.J., Rundle, S.A. and Moorcroft, J.S. 1985. Pb in playground dust and on the hands of schoolchildren. Sci. Total. Environ. 44:65-79.

Ernhart, C.B., and Greene, T. 1990. Low level lead exposure in the prenatal and early childhood periods: Language Development. Archives of Environmental Health. 342-354.

Ernhart, C.B., Morrow-Tlucak, M., and Wolf, A.W. 1988. Low level lead exposure and intelligence in the preschool years. Sci Tot. Environ. 71:454-459.

Ernhart, C.B., Landa, B., and Schell, N.B. 1981. Subclinical levels of lead and developmental deficit - a multivariate follow-up reassessment.

Farfel, M.R. and Chisolm, J.J. 1990. Health and environmental outcomes of traditional and modified practices for abatement of residential lead-based paint. American Journal of Public Health, 80, 1240 - 1245.

Feldman, R.G. 1978. Urban lead mining: Lead intoxication among deleaders. N. Engl. J. Med., 298, 1143-1145.

Fischbein, A., Anderson, K.E., Sassa, S., Lilis, R., Kon, S., Sardozi, L. and Kappas, A., 1981. Lead poisoning from do-it-yourself heat guns for removing lead-based paint: Report of two cases. Environ. Res., 24, 425-431.

Galke, W.A., Hammer, D.E., Keil, J.E. and Lawrence, S.W. 1978. Environmental determinants of lead burdens in children. In: Proceedings of an international conference on heavy metals in the environment, Toronto, Canada, 1975. Toronto: Inst. for Environ. Studies. Vol. 3, 53-74.

Hammond, P.B., Bornschein, R.L. and Succop. 1985. Dose-effect and dose response relationships for blood lead to erythrocyte protoporphyrin in young children. Environ. Res., 38, 187-196.

Harper, G.P. and Richmond, J.B. 1977. Normal and abnormal psychosocial development: In: Pediatrics, Rudolph, AM, (Ed.) 16th Ed. Appleton-Century-Crofts, New York, pp. 79-83.

Hoffnagle, G.F. 1988. Real world modelling of blood-lead from environmental sources. In Lead in Soils: Issues and Guidelines. Davies, B.E. and Wixson, B.G. (Eds.) Environmental Geochemistry and Health Series 4, Supplement to Vol 9. Science Reviews Ltd., Northwood, England. 73-94.

Inskip, M., and Atterbury, M. 1983. The legacy of lead based paint: potential hazards to do-it-yourself enthusiasts and children. Proc. Interna tional Conference, Heavy Metals in the Environment, Heidelberg, West Germany, September, 1983. CEP Consultants, Ltd., Edinburgh, UK, Vol. 1, 286-289.

Kneip, T.J., Mallon, R.P., and Harley, N. 1983. Biokinetic modelling for mammalian lead metabolism. Neurotox . 4, 189-192.

Landrigan, P.J., Gehlback, S.H., Rosenblum, B.F., Shoults, J.M., Candelaria, R.M., Barthel, W.F., Liddle, J.A. Smrek, A.L., Staehling, N.W. and Sanders, J.F. 1975. Epidemic lead absorption near an ore smelter: the role of particulate lead. N. Engl. J. Med., 292, 123-129.

Laxen, D.P.H., Raab, G.M. & Fulton, M. 1987. Children's blood lead and exposure to lead in household dust and water - a basis for an environmental standard for lead in dust. Sci. Total Environ. 66, 235-244.

Lin-Fu, J.S. 1973. Vulnerability of children to lead exposure and toxicity. N. Engl. J. Med. 289, 1229-1233, 1289-1293.

Lourie, R.S., Layman, E.M. and Millican, F.K. 1963. Why children eat things that are not food. Children 10, 143-148.

Mahaffey, K.R. 1992. Exposure to lead in childhood: the importance of prevention. N. Engl. J. Med. 327: 1308-1309.

Mahaffey, K.R., Annest, J.L., Roberts, J. and Murphy, R.S. 1982. National estimates of blood lead levels: United States 1976-1980: association with selected demographic and socioeconomic factors. N. Engl. J. Med. 307, 573-579.

Mahaffey, K.R., 1981. Nutritional factors in lead poisoning. Nutrition Reviews, 39, 353-362.

Marcus, A.H. & Cohen, J. 1988. Modelling the blood lead-soil lead relationship. In: Lead in Soil: Issues and Guidelines. Davies, B.E. and Wixson, B.G. (Eds.) Environmental Geochemistry and Health, Monograph Series 4, Supplement to Vol.9. Science Reviews Ltd. Northwood, England. 161-174.

Marino, P.E., Landrigan, P.J., Graef, J., Nussbaum, A., Bayan, G., Boch, K. and Boch, S. 1990. A case report of lead paint poisoning during renovation of a Victorian farmhouse. American Journal of Public Health, 80, 1183 - 1185.

McMichael, A.J., Baghurst, P.A., Wigg, N.R., Vimpani, G.V., Robertson, E.F. and Roberts, R.J. 1988. The Port Pirie cohort study: environmental exposure to lead and children's abilities at the age of four years. N. Engl. J. Med., 319, 468-475.

McMichael, A.J., Baghurst, P.A., Robertson, E.F., Vimpani, G.V. and Wigg, N.R. 1985. The Port Pirie study: Blood lead concentrations in early childhood. Med. J. Aust. 143:499-503.

Milar, C.R. and Mushak, P. 1982. Lead contaminated housedust: hazard, measurement and decontamination. In: Lead absorption in children: management, clinical, and environmental aspects. Chisolm, J.J. & O'Hara, D.M., (Eds). Baltimore:Urban and Schwarzenbert. 143-152.

Moffat, W.E. 1989. Blood lead determinants of a population living in a former lead mining area in southern Scotland. Environ. Geochem. Health , 11, 3-9.

Moore, M.R. and Goldberg, A. 1985. Health implications of the hemotapoietic effects of lead. 261- . In Dietary and Environmental Lead: Human Health Effects. Mahaffey, K.R (Ed.) Elsevier Biomedical Press, Amsterdam.

Moore, M.R., Richards, W.N. and Sherlock, J.G. 1985. Successful abatement of lead exposure from water supplies in the west of Scotland. Environ. Res., 38, 67-76.

National Academy of Sciences/National Research Council. 1972. Lead: Airborne Lead in Perspective. Committee on Medical and Biological Effects of Atmospheric Pollutants.National Research Council, Washington, D.c.

Needleman, H.L. 1992. (Ed.) Human Lead Exposure. CRC Press, Boca Raton, FL, USA, 286 p.

Needleman, H.L., Gunnoe, C., Leviton, A., Reed, R., Peresie, H., Maher, C. and Barrett, P. 1979. Deficits in psychologic and classroom performance of children with elevated dentine lead levels. N. Engl. J. Med. 300, 689-695.

Needleman, H. L., Schell, M. A., Bellinger, D., Leviton, A., and Allred, E. N., 1990. The Long-Term Effects of Exposure to Low Doses of Lead in Childhood: An 11-year Follow-up Report N. Engl. J. Med. 322, 83-88.

Neri, L.C., Johnson, H.L., Schmitt, N., Pagan, R.T. and Hewitt, D. 1978. Blood lead levels in children in two British Columbia communities. In: XII Annual Conference on Trace Substances in Environmental Health. Hemphill, D.D. (Ed.) University of Missouri, Columbia. 403-410.

Nordberg, G.F. (Ed.) 1976. Effects and Dose-Response Relationships of Toxic Metals. Elsevier Scientific Publishing Company, Amsterdam-Oxford, New York, pp. 7-111.

Paul, C. 1860. Study of the effect of slow lead intoxication on the product of conception. Arch. Gen. Med., 15, 513-533.

Phillips, P.E., Vornberg, D.L. and Lanzafame, J.M. 1989. Herculaneum lead study with a risk reduction analysis. In Lead in Soil: Issues and

Guidelines , Davies, B.E. and Wixson, B.G. (Eds.) Environmental Geochemistry and Health , Monograph Series 4, Vol. 9, Science Reviews Ltd., Northwood, U.K., pp. 95-104.

Piomelli, S., Seaman, C., Zullow, D., Curran, A., and Davidow, B. 1982. Threshold for lead damage to heme synthesis in urban children. Proc. Natl. Acad. Sci. Washington D.C. 79,3335-3339.

Quinn, M.J. and Delves, H.T. 1989. The UK blood lead monitoring programme, 1984-1987: results for 1986. Human Toxicol. 8, 205-220.

Rabinowitz, M.B. and Bellinger, D.C. 1988. Soil lead-blood lead relationship among Boston children. Bull. Environ., Contam. Toxicol . 41, 791-797.

Rabinowitz, M.B., Kopple, J.D. and Wetherill, G.W. 1980. Effect of food intake and fasting on gastrointestinal lead absorption in humans. Am. J. Clin. Nutr. 33:1784-1788.

Reeves, R., Kjellstrom, T., Dallow, M. and Mullins, P. 1982. Analysis of lead in blood, paint, soil and housedust for the assessment of human lead exposure in Aukland. New Zealand J. Sci. 25, 221-227.

Roberts, T.M., Hutchinson, T.C., Paciga, J., Chattopadhyay, A., Jervis, R.E., VanLoon, J. and Parkinson, D.K. 1974. Lead contamination around secondary smelters; estimation of dispersal and accumulation by humans. Science 186, 1120-1123.

Roels, H.A., Buchet, J.P., Lauwerys, R.R., Brauaux, P., Claeys- Thoreau, F., Lafontaine, A. and Verduyn, G. 1980. Exposure to lead by the oral and pulmonary routes of children living in the vicinity of a primary lead smelter. Environ. Res. 22, 81-94.

Sayre, J.W., Charney, E., Vostal, J. and Pless, I.B. 1974. House and hand dust as a potential source of childhood lead exposure. Am. J. Dis. Child. 127, 167-170.

Schmitt, N., Phillion, J.J., Larsen, A.A., Harnadek, M. and Lynch, A.J. 1979. Surface soil as a potential source of lead exposure for young children. Can. Med. Assoc. J. 121, 1474-1478.

Schroeder, S.R. 1989. Child-caregiver environmental factors related to lead exposure and I.Q. In: Lead Exposure and Child Development: An International Assessment. Smith, M.A., Grant, L.D. and Sors, A.I. (Eds.) Kluwer Academic Publishers, Dordrecht-Boston-London, pp. 166-182.

Shellshear, I.D., Jordan, D. and Hogan, D.J. 1975. Environmental lead exposure in Christchurch children: soil lead a potential hazard. N.Z. Med. 81, 382-386.

Smith, M.A., Grant, L.D. and Sors, A.I. 1989. Lead Exposure and Child Development: an International Assessment. Kluwer Academic Publishers Dordrecht-Boston-London.

Stark, A.D., Quah, R.F., Meigs, J.W. and DeLouise, E.R. 1982. The relationship of environmental lead to blood-lead levels in children. Environ. Res. 27, 372-383.

Steele, M.J., Beck,B.D., Murphy,B.L. and Strauss,H.S. 1990.   Assessing the
    contribution from lead in mining wastes to blood lead. Regulat. Toxicol.
    Pharmacol. 11: 158-190
U.S. Environmental Protection Agency. 1986. Air quality   criteria for lead.
    NC: Environmental Criteria and Assessment Office. 4 vols. (vol. 3
    particularly deals with soil and dust). EPA-600/8-83/028cF. Research
    Triangle Park.
U.S. Environmental Protection Agency. 1988. Review of the   national
    ambient air quality standards for lead: exposure analysis methodology
    and validation. Draft Report. Research Triangle Park, NC: Office of Air
    Quality Planning and Standards, Air Quality Management Division.
    August, 1988.
U.S. Dept. of Health, Education and Welfare. 1971. Medical Aspects of
    Childhood Lead Poisoning. US Dept. HEW, Health Services and Mental
    Health Administration Health Report, Washington D.C. 86: 104-143.
Walter, S.D., Yankel, A.J. and von Lindern, I.H. 1980. Age- specific risk
    factors for lead absorption in children. Arch. Environ. Health 35, **53-58.**
Watson, W.W., Witherell, L.E. and Giguerre. 1978. Increased   lead
    absorption in children of workers in a lead storage battery plant.   J.
    Occup. Med. 20, 759-761.
Yankel, A.J., von Lindern, I.P. and Walter, S.D. 1977. The   Silver Valley lead
    study: The relationship between childhood blood lead levels and
    environmental exposure. J. Air Pollut. Control Assoc. 27, 763-767.
Ziegler, E.E., Edwards, B.B., Jensen, R.L., Mahaffey, K.R. and   Fomon, S.J.
    1978. Absorption and retention of lead by infants Pediatr. Res., 12,
    29-34.

SECTION VIII Lead in soil: risk assessment and problem management.

Angle, C. and McIntire, M. 1982. Children, The Barometer of Environmental
    Lead. Adv. Pediat. 29: 3-31.
ATSDR (Agency for Toxic Substances and Disease Registry). 1988. The Nature
    and Extent of Lead Poisoning in Children in the United States: A Report
    to Congress. U.S. Department of Health and Human Services. Atlanta,
    Georgia.
Billick, I.H., Curran, A.S., and Shier, D.R. 1979. Analysis of Pediatric Blood
    Lead Levels in New York City for 1970-1976. Env. Health Perspect. 31:
    183-190.
Bornschein, R.L., Succop, P., Krafft, K., Clark, C., Peace, B., and Hammond, P.
    1986. Soil Lead, Interior House Dust Lead and Childhood Lead Exposure
    in an Urban Environment. Trace Substances in Environmental Health,
    Hemphill, D.D. (Ed.) University of Missouri, Columbia 20: 322-332.
Bornschein, R.L., Succop, P., Dietrich, K.N., Clark, C.S., Que Hee, S.S. and
    Hammond, P.B. 1985. The Influence of Social and Environmental Factors

on Dust Lead, Hand Lead, and Blood Lead in Young Children. Env. Res. 38: 108-118.

Bose, A. et al. 1983. Azarcon por Empacho-- Another Cause of Lead Toxicity. Pediatrics 72: 106.

Brunekreef, B., Noy, D., Biersteker, K., and Boleij, J. 1983. Blood Lead Levels of Dutch Children and Their Relationship to Lead in the environment. J. Air Poll. Cont. Assoc. 33: 872-876.

Carter-Pokras, O., Pirkle, J., Chavez, G., and Gunter, E. 1990. Blood Lead Levels of 4-11 Year-Old Mexican American, Puerto Rican, and Cuban Children. Public Health Reports 105: 388 - 393.

Chisolm, J.J., Jr., Mellits, E.D. and Quaskey, S.A. 1985. The Relationship Between the Level of Lead Absorption in Children and the Age, Type and Condition of Housing. Environmental Research 38: 31-45.

Clark, C.S., Bornschein, R.L., Succop, P., Que Hee, S.S., Hammond, P.B., and Peace, B. 1985. Condition and Type of Housing as an Indicatory of Potential Environmental Lead Exposure and Pediatric Blood Lead Levels. Env. Res. 38: 46 - 53.

Covello, V.T. and Allen, F. 1988. Seven Cardinal Rules of Risk Communication. U.S. Environmental Pollution Agency, Office of Policy Analysis, Washington, D.C.

Davies, B.E. and Houghton, N.J.. 1984. Distance-decline Patterns in Heavy Metal Contamination of Soils and Plants in Birmingham, England. Urban Ecology 8: 285-294.

Davies, J.C., Covello, V.T., and Allen, F.W. (Eds.) 1987. Risk Communication: Proceedings of the National Conference on Risk Communication. The Conservation Foundation, Washington, D.C.

Dietrich, K., et. al. 1987. The Neurobehavioral Effects of Early Lead Exposure. In: Toxic Substances and Mental Retardation: Neurobehavioral Toxicology and Teratology. Schroeder, A. (Ed.) Washington, D.C. AAMD Monographs. 71-98.

Elias, R.W. 1989. Soil Lead Abatement Overview: Alternatives to Soil Replacement. In Lead in Soil: Issues and Guidelines. Davies, B.E. and Wixson, B.G. (Eds) Environmental Geochemistry and Health. Monograph Series 4, Supplement to Vol. 9. Science Reviews Ltd., Northwood, U.K. 301-306.

Hallenbeck, W.H. and Cunningham, K.M. 1986. Quantitative Risk Assessment for Environmental and Occupational Health. Lewis Publishers, Inc., Chelsea, Michigan.

ICRCL (Interdepartmental Committee on the Redevelopment of Contaminated Land--Central Directorate on Environmental Pollution). 1987. Guidance on the Assessment and Redevelopment of Contaminated Land. 2nd edition. ICRCL 59/83. London.

Mahaffey, K.R. 1981. Nutritional Factors in Lead Poisoning. Nutrition Reviews 39: 353-362.

Mahaffey, K.R., Annest, J.L., Roberts, J. and Murphy, R.S. 1982. National Estimates of Blood Lead Levels: United States 1976- 1980. New England J. Med. 307: 573-579.

Mielke, H.W. 1991. Lead in Residential Soils: Background and Preliminary Results of New Orleans. Water, Air and Soil Pollution 57/58: 111-119.

Mielke, H.W., Blake, B., Burrough, S., and Hasinger, N. 1985. Urban Lead in Minneapolis: The case of the Hmong children. Envir. Res. 34 (1): 64-76.

Mielke, H.W. and Adams, J.L. 1989. Environmental lead risk in the twin cities. Centre for Urban and Regional Affairs. Minneapolis, MN.

MMWR (Mortality and Morbidity Weekly Report). 1983a. Lead Poisoning from Mexican Folk Remedies. Oct. 28, 554-555.

MMWR (Mortality and Morbidity Weekly Report). 1983b. Folk Remedy Associated Lead Poisoning in Hmong Children. Oct. 28, 555-556.

NRC (National Research Council). 1980. Lead in the Human Environment. National Academy of Sciences. Washington, D.C.325p.

NRC (National research Council). 1983. Risk Assessment in the Federal Government: Managing the Process. National Academy of Sciences, Washington, DC.

NRC (National Research Council). 1989. Improving Risk Communication. National Academy of Sciences, Washington, D.C.

Pope, A. 1986. Exposure of Children to Lead-Based Paints. PEI Associates Report to EPA, Strategies and Air Standards Division, Research Triangle Park, N.C.

Povenzano, G. 1980. The Social Costs of Excessive Lead- Exposure During Childhood. In: Low Level Lead Exposure: The Clinical Implications of Current Research. Needleman,H. (Ed.) Raven Press, NY.

Rabinowitz, M.B. and Bellinger, D.C. 1988. Soil Lead-Blood Lead Relationship Among Boston Children. Bull. Environ. Contam. Toxicol. 41: 791-797.

Reagan, P.L. and Mielke, H.W. 1991. The Monetized Costs of Childhood Pb Exposure: An Estimate for Three Minnesota Cities. In: XXV Annual Conference on Trace Substances in Environmental Health. Beck, B.D. (Ed.) Science Reviews Ltd. Northwood, England. 105-118.

Ricci, P.F. and Rowe, M.D. (Eds.) 1985. Health and Environmental Risk Assessment. Pergamon Press. N.Y.

Rice, C.A., Rischbein, A., Lillis, R., Sarkozi, L., Kon, S., and Selikoff, I.J. 1978. Lead Contamination in the Homes of Employees in Secondary Lead Smelters. Env. Res. 15: 375-380.

Rowe, W. D. 1977. An Anatomy of Risk. Wiley-Interscience, N.Y.

Szako, N.B. and Pollack, S. 1987. A Silent and Costly Epidemic: The Medical and Educational Costs of Childhood Lead Poisoning in Massachusetts. A Report by the Conservation Law Foundation of New England, Inc.

U.S. Congress. 1984. Statement by Ethyl Corporation. In: Hearing Before the Committee on Environment and Public Works United States Senate, Ninety-eighth Congress, Second Session on S. 2609, A Bill to Amend the Clean Air Act with Regard to Mobile Source Emission Contr. June 22, 1984. US Government Printing Office. Washington, D.C. p. 148.

U.S. E.P.A. (United States Environmental Protection Agency). 1985. The Costs and Benefits of Reducing Lead in Gasoline. Prepared by Schwartz, J., Pitcher, H., Levin, R., Ostro, B. and Nichols, A.L. Office of Policy Analysis.Washington D.C.

U.S. Environmental Protection Agency. 1988. Review of the national ambient air quality standards for lead: exposure analysis methodology and validation. Draft Report. Research Triangle Park, NC: Office of Air Quality Planning and Standards, Air Quality Management Division. August, 1988.

Vashistha, K. 1981. Use of Lead Tetroxide as a Folk Remedy for Gastrointestinal Illness. Morbidity and Mortality Weekly Report, Nov. 6.

Yip, R., Norris, T.N. and Anderson, A.S. 1981. Iron Status of Children with Elevated Blood-Lead Concentrations. J. Pediatrics 98: 922-925.

SUPPLEMENT I

SOIL AND BLOOD RELATIONSHIPS

In Section VII a relationship was proposed between the soil lead content and parameters affectin blood lead concentrations. The formula was :

$$S = \left(\frac{\frac{T}{G^n} - B}{\delta} \right) 1000$$

S = soil or dust guideline
T = blood lead guideline or target concentration
G = geometric standard deviation of the blood lead distribution
B = background or baseline blood lead concentration in the population from sources other than soil and dust.
n = number of standard deviations corresponding to the degree of protection required for the population at risk
δ *(delta)* is the *slope or response of the blood lead - soil(dust) lead relationship* and has the units of μg Pb dl^{-1} blood increase per 1000 μg Pb g^{-1} increment of soil or dust lead.

The expression may be solved manually (with the aid of a simple calculator) or using the following computer program. The program is written in a standard form of BASIC and assumes a printer using the Epson printer control characteristics (*e.g.*, an Epson FX dot matrix printer). Calculated values of S should usually be rounded to *e.g.*, the nearest 50 μg/g.

Computer Program

The following BASIC program performs these calculations
It was written by B. E. Davies and C. Strehlow (1992)

```
10 CLS
20 PRINT "PROGRAM TO CALCULATE TRIGGER SOIL LEAD VALUE"
30 PRINT "FROM BLOOD LEAD PARAMETERS"
40 PRINT
50 PRINT"You will need to have values ready for your current blood"
60 PRINT"lead guideline value (alert, regulatory value) in ug/dL"
70 PRINT
80 PRINT"You also need a value (ug/dL) for your background blood level"
90 PRINT"e.g., the lead value from a local control group"
100 PRINT
110 PRINT"Also, values for blood Pb geometric deviation and delta, the"
```

```
120 PRINT"the slope of the soil(dust)-blood response curve"
122 PRINT
124 PRINT "PRINTER must be switched ON and READY"
130 PRINT
140 PRINT"DO YOU WANT TO CONTINUE? y/n":INPUT QQ$
150 IF QQ$ = "y" THEN 180
160 IF QQ$ = "Y" THEN 180
170 SYSTEM
180 REM
190 LPRINT "PROGRAM TO CALCULATE TRIGGER SOIL LEAD
    VALUES"
200 LPRINT "FROM BLOOD LEAD PARAMETERS"
210 LPRINT
220 LPRINT"DATE OF RUN    = ":LPRINT DATE$
230 LPRINT"TIME OF RUN    = ":LPRINT TIME$
240 PRINT
250 REM
260 LPRINT
270 PRINT
280 PRINT
290 PRINT"INPUT YOUR BLOOD LEAD GUIDELINE VALUE (T)"
300 INPUT T
310 LPRINT"BLOOD GUIDELINE VALUE (T)        = ":LPRINT T
320 PRINT"INPUT YOUR BASELINE/BACKGROUND BLOOD LEAD (B)"
330 INPUT B
340 LPRINT"BACKGROUND BLOOD LEAD (B)        = ":LPRINT B
350 PRINT"INPUT YOUR GEOMETRIC DEVIATION OF THE BLOOD
    LEAD DISTRIBUTION"
360 PRINT"THIS IS TYPICALLY IN THE RANGE 1.3 TO 1.5"
370 INPUT G
380 LPRINT"BLOOD LEAD GEOMETRIC DEVIATION (G)  = ":LPRINT G
390 PRINT"INPUT DELTA, THE SLOPE OF THE RESPONSE CURVE"
400 INPUT DELTA
410 LPRINT"CHOSEN VALUE OF DELTA            =":LPRINT DELTA
420 PRINT"INPUT N, THE NUMBER OF STANDARD DEVIATIONS TO
    PROTECT "
430 PRINT"THE REQUIRED PROPORTION OF THE ION"
440 PRINT"95% = 1.64, 98% = 2.05, 99% = 2.32"
450 INPUT N
460 LPRINT"CHOSEN VALUE OF N                = ":LPRINT N
470 S=(((T/G^N)-B)/DELTA)*1000
471 REM if S , round to nearest 10
472 IF S THEN S=(INT((S+5)/10))*10
473 REM if s1000, round to nearest 25
474 IF S=995 THEN S=(INT((S+12.5)/25))*25
475 REM round to 50 would be  s=(int((s+25)/50))*50
```

476 REM round to 100 would be s=(int((s+50)/100))*100
480 PRINT S
490 LPRINT
500 LPRINT"COMPUTED SOIL GUIDELINE VALUE   = ";S;"
    MG/KG(ppm)"
510 REM deleted 2nd lprint s & changed : to ; in line 500
515 IF S THEN 575
520 PRINT"DO YOU WANT MORE CALCULATIONS? Y/N"
530 INPUT QQ$
540 IF QQ$ = "Y" THEN 250
550 IF QQ$ = "y" THEN 250
560 LPRINT "END OF RUN AT ":LPRINT TIME$
570 SYSTEM
575 REM caution if negative value
577 PRINT "NEGATIVE value. READ message on PRINTER"
580 LPRINT "Negative value means that Baseline PbB or degree of protection"
590 LPRINT " is too high to allow ANY contribution from soil lead."
600 LPRINT " Try varying input assumptions"
610 GOTO 520
620 END

Examples of Calculations

The reader should note that the computer program has been written to provide a rounding of the numerical result for S in order to avoid a spurious precision. The following are *manual* calculations.

Example 1

$T = 15 \ \mu g \ dl^{-1}$

$n = 2.05$ for 98% of population less than T

$G = 1.40$

$B = 4 \ g \ dl^{-1}$, no other significant sources

$\delta = 2 \ \mu g \ dl^{-1}$ per 1000 g/g

$S \ = 1763$ (rounded, 1800) g Pb g^{-1}

(Note that if T is enetered as 10, the new CDC limit, then S = 500)

Example 2

as above, except take $\delta = 5$ µg dl^{-1} per 1000 g g^{-1}

$\quad$ S $=$ 705 µg Pb g^{-1}

Example 3

as in (1), except air lead contributes an additional 3 µg dl^{-1} above background to the baseline blood lead concentration. *i.e.,* B $= 4+3$

$\quad$ S $= 263$ µg Pb g^{-1}

This level of (rounded) 250 µg g^{-1} could be used, if no measures were taken to reduce air lead exposure. If such measures were taken, an appropriate S between 250 and 1,800 µg g^{-1} would be recalculated.

Example 4

as in (1), except n $= 3.04$; *i.e.,* for 99.9% of population to have a blood content below T $= 15$ µg dl^{-1}.

$\quad$ S $= 697$ µg Pb g^{-1}

Example 5

it is desired to set a guideline for undeveloped land using the values as in (1), but with n $= 0$ for the mean blood lead content to be below T $= 15$ µg dl^{-1}.

$\quad$ (Note: $1.4^0 = 1.0$)

$\quad$ S $=$ 5,500 µg Pb g^{-1}

Example 6

given the guideline of 1,800 µg g^{-1} found in (1), it is desired to set a guideline for undeveloped land at 2 standard deviations above the soil mean, using a soil lead geometric standard deviation of 2.0

$\quad$ S $=$ 7,200 µg Pb g^{-1}

Example 7

as in (6) above, but taking the mean of 700 µg g^{-1} found in (2) as a starting point

S = 2,800 µg Pb g^{-1}

Example 8

T = 25 µg dl^{-1}

n = 3.04, for 99.9% of population less than T

G = 1.43

B = 5 µg dl^{-1}

δ = 2 µg dl^{-1} per 1,000 µg Pb g^{-1}

S = 1,714 µg Pb g^{-1}

Example 9
as in (8), except n = 3.71 for 99.99% of population less than T (10 out of every 100,000 at risk)

S = 816 µg Pb g^{-1}

Example 10
given the guideline of 800 µg Pb g^{-1} found in (2), it is desired to set a guideline for amenity grassland at 2 standard deviations above the soil mean, using a geometric standard deviation of 2.10.

S = 3,528 µg Pb g^{-1}

Example 11
it is desired to set a guideline for the amenity grassland using the values as in (8), but with n = 0 for the mean blood lead to be below T

S = 10,000 µg Pb g^{-1}

SUPPLEMENT II

SOIL SAMPLING AND ANALYSIS

There is no international, standardised method for soil sampling although there is broad agreement between different workers in different countries on how it should be done. Most early sampling was done for agricultural purposes particularly for providing soil fertiliser advice. In recent decades geographers, ecologists, geochemists and environmental scientists have also been involved in soil sampling for their own purposes. Therefore, while most researchers use broadly the same strategies and techniques there is little consistency in detail. All will agree that the sample must be "representative". A useful rule of thumb is that 1 hectare of soil to 15 cm depth weighs 2.8×10^5 kg or 1 acre to six inches depth weighs 2.5×10^6 lb. A typical bulked soil sample will amount to only 1 - 2 lb material or, approximately, up to only 1 kg. Clearly it is not easy to ensure a representative sample and basic research concerning the lateral heterogeneity of soil at different scales has not been done. There is little agreement on the depth of soil to be sampled for surface samples, some favour 10 cm others 15 cm. A variety of tools is used including spades, trowels, screw and pipe augers. For sub-surface horizons pedologists will favour genetic horizons whereas environmental scientists may prefer standard depths, typically 50 cm. Supplement IIA is very specific advice used in USA Superfund work. Supplement IIB is based on general guidance given to workers in the Environmental Science Department of Bradford University

SUPPLEMENT IIA

SUPERFUND SOIL LEAD ABATEMENT DEMONSTRATION PROJECT

PROTOCOL FOR SOIL SAMPLING AND ANALYSIS

SOIL SAMPLING

A. Site Description

1. General Site Description

For each location, a detailed drawing should be made that shows the boundary of the lot, the position of the main building and any other buildings such as storage sheds or garages, the position of the sidewalks, driveways, and other paved areas, the position of the play areas if obvious, and the position of the areas with exposed soil (grassy or bare). Show down spouts and general drainage patterns. Identify each soil subarea by letter or number. If a large soil area needs to be divided into smaller patches for sampling convenience, show how this division was made.

In addition to the diagram, briefly describe the location, including the following information:

> Type of building construction
> Condition of main building
> Condition of lot (debris, standing water, vegetation cover)
> Nature of adjacent property
> Presence and type of fence
> Animals on property
> Apparent use of yard (toys, sandbox, children present)
> Underground utilities

2. Subarea description

For each soil subarea identified on the general diagram, draw a full page diagram showing the approximate dimensions and position relative to the building foundation. Indicate vegetation and bare soil areas, as well as obvious traffic patterns. Identify the category of landuse, such as roadside, property boundary, adjacent to foundation, play area. Select an appropriate sampling scheme and mark the sample locations on the diagram.

3. Sampling Schemes

The sample scheme selected for each subarea must adequately characterize the potential exposure of children to lead in the dust from this soil. It must identify the areas of high lead concentrations, and the general distribution pattern of lead concentrations at the soil surface. For abatement purposes, the depth to which lead has penetrated the soil profile must be determined. Consequently, selecting the most appropriate sampling scheme is the critical element in the site description.

Several options are offered for the best judgement of the investigator.

a. Line Source Pattern

This pattern can be used whenever the source of the lead is thought to be linear, such as along a building foundation, a fence row, a street, or beside a garage. Draw a line parallel to the source, such as the foundation of the main building, approximately 0.5 metres (16 inches) from the foundation. Repeat at the property boundary if the subplot is more than three meters wide (10 ft.), and add a third parallel line between the first two if the subarea exceeds 5 metres (16 ft.) in width. Divide each line into segments that do not exceed 7 metres (23 ft.) in length. Take one composite of 5-10 cores along each line segment. A subarea, for example, that is at the side of the main building and measures 12 x 7 metres would have three lines of two segments each. The lines would be parallel and approximately 3 metres apart. They would be 12 metres long and consist of two 6 metre segments each, making a total of six samples, each being a composite of at least five cores divided into a top 2 cm sample and a bottom 2 cm sample. (Figure 3A).

b. Targeted Pattern

This method is intended to be used in conjunction with the line source or grid patterns as a means of sampling obvious areas that would be missed by the regular patterns. In using the targeted pattern, the investigator should select those locations within the subarea that are likely to reflect potential exposure to lead in soil dust. These may be play areas, paths, drainage collection areas, of areas that are likely to contribute dust to other surfaces that children use. Determine the number of samples to be taken by identifying distinctive landuse characteristics (path, swingset, sandbox), and take a composite of 5-10 cores for each sample. (Figure 3B)

c. Small Area Pattern

When the subarea is less than two meters in each dimension, or when the accessible area of a larger plot is less than four square meters, a single composited sample may be taken if it appears that such a sample would adequately represent the subarea. (Figure 3C).

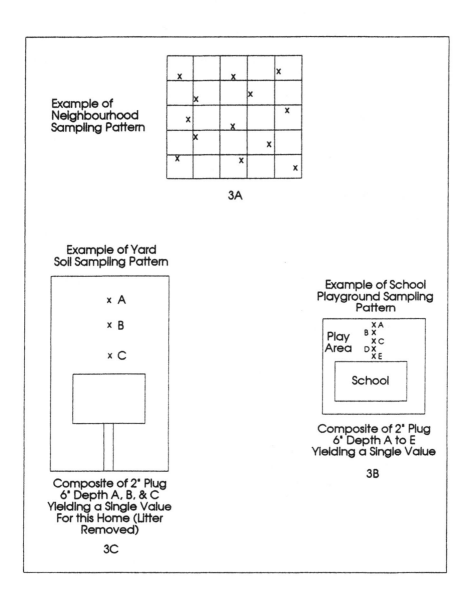

Figure 3 *Preliminary soil sampling designs*

d. Grid Pattern

Establish a rectangular grid of intersecting lines 2 – 10 metres apart, and sample each rectangular area. For larger areas, randomly select the rectangles to be sampled. In each rectangular area, mark three lines parallel to the longest axis, and composite 5–10 cores along each line. Since the rectangle sould not exceed four meters, there is no need to divide the line into segments. Therefore, each rectangle should have six samples of 5–10 composites each. Use this pattern when the subarea is generally uniform and there is no reason to suspect large variations in lead concentrations (Figure 4A).

e. Visual Location.

When the sample sites have been located on the subarea diagram and the sample collection is ready to proceed, locate each sample with a flag and visually confirm an even and representative distribution of sample location (Figure 4B).

B. Sample Collection

The flags or other markers represent the center of the sample location for the targeted and small area patterns. For the line source and grid patterns, the flags indicate the sampling lines. Take at least five but not more than ten cores randomly selected from within the sampling area of the targeted and small area sampling patterns, and uniformly spaced along the sampling lines of the line source and grid patterns. The cores make a composite identified as a single sample. A sample record sheet is used to record information about the composite. The corer should be clean and free of lead contamination. Vegetation and debris can be removed at the point of insertion, but do not remove any soil or decayed litter. The corer should be driven into the ground to a depth of at least 10 cm, 15 cm if possible. If the 10 cm depth cannot be reached, the corer should be extracted and cleaned, and another attempt made nearby. If the second attempt does not permit a 10 cm core, the sample should be taken as deep as possible, and the maximum depth of penetration noted on the sample record sheet. Every effort should be made to take all cores of a composited sample at the same depth.

The cores of each plot should be examined for debris, artifacts, and any other evidence of recent soil disturbance. These should be noted on the subarea description sheet, as should a brief description of the soil color and soil type.

For each sample location, the top 2 inches or 5 cm segment of each of the cores are composited into one sample, and the bottom 2 to 6 inches or 5 to 15 cm segment combined into a second (Figure 4C). For the surface segment, debris and leafy vegetation should not be included with the sample. However, no soil or decomposed litter should be removed, as this is the most critical part of the soil sample and is likely to be the highest in lead concentration.

The soil core segments should be composited in sealable polyethylene containers suitable for prevention of contamination and loss of the sample. The sample identification number should be placed on the container and the sample

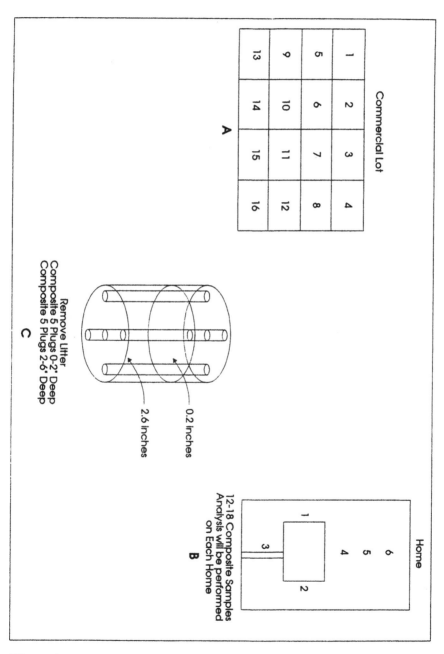

Figure 4 *Detailed soil sampling*

record sheet. After each sample composite, the corer should be cleaned by reinsertion in the next sampling area. Store the composited soil sample at ambient temperature until returned to the laboratory

C. Sample Handling And Storage

The sample containers should be sealed to prevent loss or contamination of the sample. Shipping containers should also be airtight. Storage should be in a cool, dry location.

D. Record-Keeping And Sample Custody

Soil sample records for each location consist of a location diagram and description, a plot diagram for each distinct soil plot, and sample record sheet for each sample in a plot. The sample record sheets should also contain space for chain-of-ustody documentation.

Samples should be sequentially numbered within each subarea. Each location diagram, subarea description, and sample record sheet should bear all sample numbers and the signature of the person responsible for verifying the quality of the information collected. This signature certifies that there has been no misuse of the sample protocol, no mistake in recording the information, and that the information is sufficient to clearly identify these samples for comparison with other types of samples taken at the same location, such as street dust, house dust, house paint, blood, and hand dust. These documents also establish the chain of custody required for the Quality Assurance Plan.

When the sample is delivered to the laboratory, custody is relinquished by the field investigator and received by the lab supervisor by signatures on the sample record form.

SAMPLE ANALYSIS

A. Method Of Analysis

Three methods of analysis have been considered. They are Atomic Absorption Spectroscopy (AAS), Inductively Coupled Plasma Emission Spectroscopy (ICP), and X-Ray Fluorescence (XRF). The XRF method is the approved method for routine analyses, whereas the AAS method should be used for standardization.

1. Sample Definition

The representative urban soil sample is defined as the soil from 0-2 inches (0 - 5 cm) depth that passes a 250 cm stainless steel seive. This fraction is comprised of small particles, and the concentration of lead believed to be closely related to that of particles on the hands of children. The fraction is also homogeneous enough to allow reliable analysis by X-Ray fluorencence.

2. Sample Preparation

Sample preparation requires that the sample be air dried and separated by particle size before being digested by wet chemistry. Drying is done at room temperature overnight or until the sample can be easily disaggregated by hand or with a rolling pin. The full sample should be brought to complete disaggregation by passing through a 2 mm sieve, using the fingers or a stainless steel tool to crush the larger soil particles. Material larger than 2 cm should be discarded. Soil should not be milled to a fine powder with a morter and pestle or any other grinding device.

The fraction that passes the 2 mm sieve is now called the total soil fraction. A portion of this sample is retained for possible reference analysis, but the larger fraction is passed through a #60 mesh sieve (250 μm), giving a fine soil fraction identified as the "Urban Soil Sample". The portion that does not pass the #60 mesh sieve should be discarded, as only the total soil fraction (mm) and the fine soil fraction will be analyzed.

About 5–10% of the retained total soil samples should be analyzed. An aliquot is ground so that it all passes a #60 mesh (250 μm) sieve, mixed well and analyzed. Grinding is necessary to provide low/appropriate variance in XRF analysis.

During the processing of the sample, it should be remembered that small soil particles may individually be as high as 50,000 μg Pbg^{-1}, and paint fragments as high as 300,000 μg g^{-1}. Care should be taken to clean equipment between samples. The sieves may be cleaned by tapping on a hard surface to remove residual particles, or any other dry method. Wet washing is not recommended as this will interfere with the size calibration.

Care should also be taken thoroughly to homogenize the separated sample before removing the aliquot for analysis. Shaking will cause separation. Tumbling or stirring is recommended.

B. Atomic Absorption Spectroscopy (To be used for primary standards)

1. Wet Digestion

The extraction procedure used for solubilizing soil lead is critical to the interpretation of the results of the Superfund Soil Lead Abatement Demonstration Projects. Even in the absence of analytical errors, the data may not represent the same lead concentrations from sample to sample unless the correct extraction procedure is used. The method selected here does not represent the total extraction of lead, but the breakdown of the organic material and the leaching of lead from the inorganic soil fraction. The methods measure total non-matrix soil lead, because no other extractable fraction has been experimentally shown to measure bioavailable, or non-HF extractable, soil lead. Hot HNO_3 has been repeatedly shown to extract total non-matrix soil lead, or at least 95% of soil lead, compared to a total soil dissolution method (HF). The 1.0 N HNO_3 cold shake method has been shown to extract as much lead as the hot HNO_3 extract, except for unpolluted soils where a higher fraction of the total soil lead is within the matrix of soil particles.

The sample should be oven dried at 105°C for 24 hours or until a constant weight is achieved. The aliquot should be placed in a 150 ml beaker and covered with a watch glass. Class A borosilicate glassware and stainless steel tools should be used throughout the sample processing. Low density conventional polyethylene containers may be used to store the solution prior to analysis.

An aliquot of 1 g soil is normally considered representative of the whole sample if the soil is well mixed. Prior to removing the aliquot, the sample should be stirred with a spatula or rod. Shaking the container can cause the sample to separate by particle size.

2.   Hot HNO₃ Extraction

Add 50 ml 7N HNO_3, cover and digest gently at 95°C for 2 hours, stirring occasionally. If excessive foaming occurs, remove from the heat periodically until foaming subsides. Maintain at least 25 ml in the beaker by adding 7N HNO_3 as necessary.

Cool and dilute with 10 ml 1N HNO_3. Filter through Whatman No. 42 filter paper into a volumetric flask. Rinse filter and labware with 1N HNO_3, and dilute to volume.

3.   Cold HNO₃ Extraction

Weigh the 1 g aliquot into a 4 oz. cup. Add 50 ml 1.0 N HNO_3 to each cup. Screw the lid on tightly and place on a reciprocal shaker. Adjust the speed of the shaker to maintain a suspension of the soil particles. Shake for one hour, then filter. Rinse with 1.0 N HNO_3. Dilute to standard volume.

4.   Analysis

Analysis by flame AAS should be at 283.3 nm, with background correction. Working standards should be prepared fresh daily, in the range of 2-50 $\mu g\ g^{-1}$, in a 1.0 N HNO_3 matrix.

5.   XRF Analysis

Approximately 2 g of loose soil sample are poured into sample cups (Somar Labs, Inc., Cat No. 340), fitted with windows of 1/4 mil thick X-ray polypropylene film (Chemplex Industries, Inc., Cat NO. 425). The sample cup should be at least half full. The sample cup is sealed with a sheet of microporous film (Spex Industries, Inc., Cat No. 352A) held in place by the snap-on sample cup cap. The exact weight of the sample is not important, but should be in the range of 2-6 g.

The instrument configuration for the Kevex Delta Analyst Energy Dispersive X-ray Spectrometer is:
1)   Kevex Analyst 770 Excitation/Detection Subsystem:
a) X-ray tube: Kevex high output sodium anode
b) Power supply: Kevex 60 kV, 3.3 mA
c) Detector/cryostat: Kevex Quantum - UTW lithium, drifted silicon. 165 eV
    FWHM resolution at 5.9 KeV.

2)   Kevex Delta Analyzer:
a) Computer mainframe: Digital Equipment Corp., PDP 11/73
b) Computer software: Kevex XRF Toolbox II, Version 4.14
c) Disk drives: Iomega Bernoulli box, dual drives, 10 MB
d) Pulse processor: Kevex 4460
e) Energy to digital converter: Kevex 5230

3)   Operating condition
a) Excitation mode: Mo secondary target with 4mil thick Mo filter
b) Excitation conditions: 30 kV, 1.60 mA
c) Acquisition time: 300 livetime seconds
d) Shaping time constant: 7.5 microseconds
e) Sample chamber atmosphere: air
f) Detector collimator: Tantalum

4)   Analytical conditions:
a) Escape peaks, but not background should beremoved from all spectra
b) The intensity ratio, defined as the integral of counts in the Pb (LA) window
    divided by the integral of the counts in the Mo (KA) Compton scatter
    window, should be determined for each spectrum
c) The intensity ratios for the standards should be used to determine a linear least
    squares calibration curve The acquisition time (3c) may be reduced at the
    discretion of the lab supervisor.

6.   *QA/QC*.
By blind insertion into the sample stream (where possible), the QA/QC officer
will provide the following blanks at the indicated frequency. At the discretion of
the project director, the field team will collect one blank per day by carrying a
sample of clean quartz sand into the field in a normal sample container. The
sample container will be opened and exposed during the collection of one sample,
then closed and returned to the lab. The field blank can be split into two aliquots.
One aliquot, the field blank, can be analyzed directly with no further treatment.
The second aliquot (the sample blank) can be analyzed after it has passed through
the sample stream (except sieving). The field blank represents contamination
added in the field, the sample blank represents contamination added in the field
and during storage and sample preparaion.

A project standard soil sample will be prepared and distributed at the
beginning of the study. This will be used as a laboratory control. For XRF
analysis, there is no need for a reagent blank.

Sample blank	1/field sampling day
Lab control	1/20 samples
Reagent blank	3/reagent batch

Additionally, split sample (duplicate) analyses and spiked samples** will be determined as follows:

Split soil	1/20 samples
Spiked soil	1/20 samples

The spiked soil samples will be prepared by mixing dried and sieved soil of known concentration with the sample. Spiked soil samples may be used at the discretion of the project director. Additional split soil samples will be sent to a designated QA/QC laboratory for analysis using the hot HNO3 method, one for each 40 samples.

An interlaboratory comparison, similar to the soil pilot study, will be conducted during each six month period, with 10-20 samples from each laboratory, including the QA/QC lab. These samples will be dried, but not sieved.

SUPPLEMENT IIB

Disclaimer

These Technical Leaflets are published only for guidance. Neither B.E.Davies nor the University of Bradford can accept any responsibility for any consequences of their use.

UNIVERSITY OF BRADFORD
DEPARTMENT OF ENVIRONMENTAL SCIENCE
LABORATORY TECHNICAL LEAFLET NO. 1.1
COLLECTION OF SOIL SAMPLES

Soil samples may be from horizons of a profile (profile samples) or a bulking of auger cores taken to be representative of a sample area (field samples). The required apparatus is common to both kinds of sample.

Samples are collected using a screw or pipe auger (for field samples) or spades and trowels (profile samples). These tools are best made of mild steel and they should be clean and free of oils, paint and corrosion. Stainless steel is suitable for many purposes but the possibility of contaminating the sample with nickel and chromium should be kept in mild. The sample is put into clear polyethylene bag and enough soil to half fill a bag 20 x 30 cm is usually sufficient. The most convenient way to label a sample is using a cloakroom ticket but do not put it in with the soil. Rather, fold the bag and soil into a packet, insert the label masking underneath a fold of the plastic and secure with tape: the identifying label may now be read from outside. Alternatively, and generally better, put the bag into a second one and seal. Self-sealing bags may be used. Cloakroom tickets are normally sold in pairs so the other ticket may be used to form part of the field record.

FIELD SAMPLES

A sample of soil weighing 1 kg when dry, collected from one acre (0.4ha) to a depth of 15 cm, represents about 2 million kg of soil in the field – so great care must be taken if the sample is to be representative.

Most sampling for advisory purposes, concerned with the manurial or lime requirements of soil, is carried out between the harvesting of one crop and the sowing of the next. such sampling should, if possible, precede ploughing or other cultivations. In general, sampling should never follow soon after liming or manuring. Full information on the site data, land use, the previous and present crops, cultivation, manuring etc., should be recorded.

A preliminary brief survey of the field should be made to assess the degree of

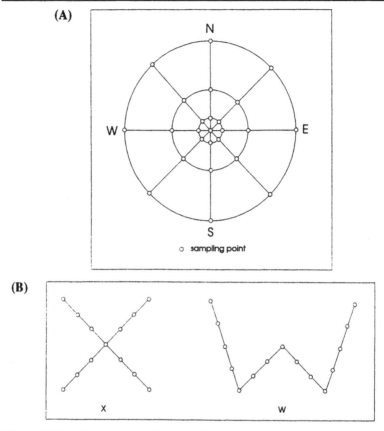

Figure 5 *Sampling patterns for agricultural soils. A, Circular grid for the survey of suspect areas. B, Non-systematic patterns.*

uniformity of the soil. Crop growth may give some indication and the auger is very useful for detecting differences in profile characteristics, texture, drainage and depth of soil. Areas showing marked differences should be sampled separately.

The soil sample for laboratory examination should be a thorough mixture of equal amounts collected from a number of random points within the area. The number of sampling points will depend on the variation in the area and the precision required, but, for fields not exceeding 4 ha (10 acres) not less than 25 auger samples should be taken. If fields are larger than 4 ha it is better to subdivide them, taking bulked samples for each smaller area.

Atypical areas are avoided : headlands, where fertilizer bags may have been dumped, gateways, areas around trees and feeding troughs, areas where the soil itself differs etc. A W-shaped path should be followed across the sampling area.

On cultivated land the sample should be taken to the depth of ploughing or digging, normally 15-20 cm. On grassland the upper 8 cm may be more

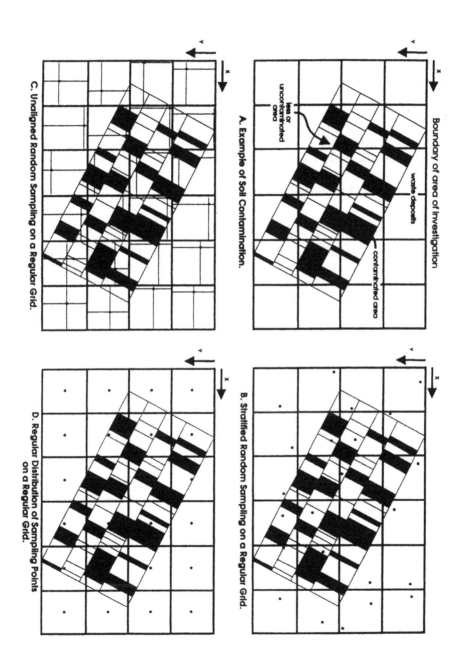

Figure 6 Sampling pattern for contaminated soils

appropriate, although if permanent grassland is to be brought into cultivation two samples, 0-8 and 8-15 cm may be taken.

An alternative method may be used when sampling is to be done at some predetermined locality, *e.g.*, when sampling along a transect or on a grid basis. If the required sample point is not available (for instance, it falls on a road or in a river, *etc.*) follow some consistent rule of thumb to transfer it. You might always move, *e.g.*, 100 m due north and sample, or, if still unsuitable, 100 m due east of the original point, then south and west – until a sample can be collected. At the point, sample around the circumference of a circle, of radius 1 m from the point, taking cores at 0°, 45°, 90°, 135°, 180° *etc.* and one from the centre. Repeat around a second circle 2m from the centre. Coring positions and radius can obviously be altered according to need.

Figures 5 and 6 illustrate sampling strategies.

PROFILE SAMPLES

Location of the profile pit is a matter for personal judgement. The pit should be started so that, when finished, one face will be fully in the light : a shadow across a face will spoil the photographic record. It helps, when filling in the pit to shovel the soil on to a large tarpaulin or plastic sheet. Cut the turfs carefully so that the field may be restored as near as possible to its initial state. After photography describe the profile and decide whence the samples should be taken. Sample the deepest horizons first, and collect ample material for all the analyses envisaged, and collect enough to ensure that the sample is truly representative of the horizon. The sampling can be done with a sharp nosed builder's trowel, or with a laboratory spatula. It is useful to work with two trowels, one to free the sample and the other to catch it. Assemble the bags of soil from right to left and, subsequently, number them left to right : the lowest number will now correspond to the uppermost sample.

HEALTH AND SAFETY NOTE

When working outside remember to wear appropriate clothing. In rugged terrain carry spare clothing, maps, compass and whistle. Always obey the country code.

B. E. DAVIES (1986)
UNIVERSITY OF BRADFORD
DEPARTMENT OF ENVIRONMENTAL SCIENCE
LABORATORY TECHNICAL LEAFLET NO. 2.1
PREPARATION OF SOIL SAMPLES

On arrival at the laboratory the sample must ordinarily be registered in the departmental collection. The field number is replaced by the permanent sample number and the following details recorded in the record book : grid reference (full reference), date, abbreviated address, sample type (eg, field or garden, sample depth) any other pertinent information and name of sampler.

If it is intended to determine an unusually labile constituent (eg, moisture content, nitrate/ammonia nitrogen, mercury, pH of waterlogged soils etc.) the soil is analysed immediately after it arrives in the laboratory. The only preparation possible will be the removal of stones and large biological debris. For all other determinations the sample is 'air-dried' before analysis. The soil is spread thinly on a clean polyethylene sheet and either allowed to dry on the bench-top at room temperature, or in an oven at approximately 25°C. In either case the soil must not be exposed to aerosol contamination or to drips from condensing moisture. It is best not to let clay soils dry out completely since they tend to harden and the subsequent grinding is made difficult.

When dry the soil sample is gently ground in an acid-clean porcelain pestle and mortar. Prolonged grinding should be avoided and it is not necessary to break down all aggregates. The sample is sieved through nylon mesh of 2mm aperture to yield 'fine earth' which is then stored in a polyethylene bag. Coarse (>2mm) material may be washed for further sieving and identification of rock type. If the coarse material is likely to exceed 10% of the total sample weight then the actual proportion should be estimated by weighing and subsequent calculations based on the fine earth should be adjusted. Thoroughly clean the pestle and mortar between samples.

The polyethylene bag containing the dried sample should be stored in numbered boxes. Close the bag in the same way as when it was filled in the field. Secure with masking tape since rubber bands may perish.

HEALTH AND SAFETY NOTE

Face masks should be used to reduce the inhalation of dust. Urban soils, especially parks, may contain the eggs etc. of harmful organisms so wash your hands carefully.

B E DAVIES (1988)
UNIVERSITY OF BRADFORD
DEPARTMENT OF ENVIRONMENTAL SCIENCE
LABORATORY TECHNICAL LEAFLET NO. 4.1
SOIL EXTRACTION : HEAVY METALS (Beaker Method)

This method extracts approximately 100% total Zn, 80% total Pb and 60-80% total copper from the soil.

Weigh approximately 5g fine earth into a labelled and weighed 50 ml beaker. Put to dry at 100°C for 24 hours; reweigh; ignite at 430°C for 24 hours, cool, reweigh. Record all weights (W1, W2, W3, W4).

To the soil add concentrated nitric acid (20 cc), cover with a glass and heat on the hotplate (110°C) for 30 minutes. Remove and rinse cover glass and take the contents to near dryness (@120°C). The soil is "dry" when it does not move on tilting the beaker but the soil surface may gleam slightly : baking MUST be avoided. Repeat this procedure once.

The dry soil slowly add 0.01 M HNO_3 (normally 50 ml but 25 ml if the total Pb is 20 ppm); break up the soil and stir with a glass rod; rinse the rod with the last few mls of acid. Set to warm on the hotplate (50°C) for 15 minutes, then filter through a fluted Whatman no. 540 paper into 25 ml volumetric flasks. Make up to the mark with distilled water.

SAFETY NOTE

In all operations involving concentrated nitric acid, wear safety goggles and gloves. Do not wear sandals or "open" footwear. Wash acid splashes immediately with copious cold water.

CALCULATIONS

Moisture content should be expressed as % oven dry weight. The after ignition is a measure of the soil's organic content and should also be expressed as % oven dry weight. A computer program for carrying this out is available.

B E DAVIES (1988)
UNIVERSITY OF BRADFORD
DEPARTMENT OF ENVIRONMENTAL SCIENCE
LABORATORY TECHNICAL LEAFLET NO. 2.5
ANALYTICAL ERRORS AND ANALYTICAL RESULTS

This leaflet is concerned with the errors which arise during laboratory analysis. It is assumed the reader is familiar with simple statistical techniques and any standard text should be consulted for the formal treatment and explanation of terms used here. One useful book is R.E. Parker's 'Introductory Statistics for Biology', Studies in Biology No. 43 (1979) published by Edward Arnold.

Laboratory measurements cannot yield the true value of any property, all that can be achieved is a best estimation for the given circumstances of analytical skill, standard of instrumentation etc. Two fundamental concepts are ACCURACY and PRECISION. A result is said to be *accurate* if it is close to or the same as the true result even though the latter is never actually known. There are several ways to check accuracy. Laboratories claiming to work to high standards should maintain a collection of *reference samples* which are analysed in every batch to monitor the laboratory's analytical quality. These reference samples are purchased from bodies which *certify* them after many analyses by many laboratories by different analytical methods. Two popular sources of reference samples are the U.S. Bureau of Standards (now US National Institute for Standards Technology) and the European Community (BCR) for soils, sediments and plant materials. It is also usual to maintain an in-house reference material which is included in every analytical batch and it is 'certified' by comparison with bought-in samples and by circulation for analysis to cooperating laboratories. It is common for laboratories to cooperate by analysing 'round robin' samples. *Precision* is the reliability of the measurements and is checked by replication and reanalysis.

The causes of inaccuracy and imprecision lie in laboratory errors. SYSTEMATIC ERRORS are those which consistently cause a result to be wrong in the same direction and they are amenable to identification and compensation. An example is when a precipitate is transferred from a filter to a beaker for a small amount tends always to be left behind. RANDOM ERRORS are the main cause of imprecision. Examples are when flame conditions vary during the use of a flame photometer or if a pipette is occasionally held incorrectly and therefore fails to deliver the designated volume.

It is usual to replicate analyses so as to minimise random errors and improve precision. Replicates of the original sample should be dispensed and, in turn, the analyses made on each sub-sample should be replicated. The data are treated standard statistical techniques.

The MEAN ($\overline{X}$) result is normally reported. The variability of the replicates is expressed by first calculating the STANDARD DEVIATION (s) for N results and then deriving the STANDARD ERROR (E) when E= S /√N.) The result is tabulated as

$$X \pm E \quad \text{or} \quad X \pm S$$

and for practical purposes the 95% confidence level is assessed from twice the standard error (formally, $X \pm 1.96$ (s^2 /N). Another useful measurement is the % COEFFICIENT OF VARIATION (CV) when

$$CV = 100s/\overline{X}$$

(Some analysts use the RELATIVE DEVIATION = $s/\overline{X}$). As a rule of thumb the precision or the replicated analyses for one sub-sample should be more than 5% and that for the whole sample should not exceed 10%. It is rarely possible to quantify the field sampling error.

Duplication is the most common laboratory replication when the formula for the standard deviation simplifies to

$$s = \frac{a-b}{\sqrt{2}} \quad \text{for two results a, b.}$$

Finally, a common problem with instrumental analysis is that the reading for a sample cannot be distinguished from that of the instrumental blank. Such a result must never be reported as zero but as detection limit which can be estimated *crudely* as the analyte concentration corresponding to 1% transmission. In general, do not tabulate values as 'blank', or 'n.d.' if no result was available.

Printed and bound by CPI Group (UK) Ltd, Croydon, CR0 4YY

18/10/2024

01776265-0003